Memoirs of the American Mathematical Society

Number 190

Robert J. Sacker and George R. Sell

Lifting properties in skew-product flows with applications to differential equations

Published by the

AMERICAN MATHEMATICAL SOCIETY

Providence, Rhode Island

VOLUME 11 · NUMBER 190 (End of volume) · JULY 1977

Acknowledgements

The research on this paper was begun while the authors were visiting the Istituto Matematico dell' Università di Firenze under the auspices of the Italian Research Council (C.N.R.) in 1972. Partial support for this research was also given to R.J.S. by U.S. Army Grant DA-ARO-D-31-124-71-G176 and to G.R.S. by NSF Grant GP-38955 and NSF Grant MCS 76-06003.

We would like to express our sincere appreciation to Miss Peggy Gendron and Mr. Leon Lemons for their accurate typing of the manuscript.

AMS(MOS) subject classifications(1970) Primary 34C35, 54H20, 58F10 . Secondary 34D20, 34K20.

Key words and phrases: Almost periodicity, covering space, distal flow, equi-continuous flow, lifting properties, ordinary differential equations, retarded differential equations, scalar equations, separatedness, skew-product flows, triangularization.

Library of Congress Cataloging in Publication Data CIP

Sacker, Robert J 1937-
Lifting properties in skew-product flows.

(Memoirs of the American Mathematical Society ; no. 190)
Bibliography: p.
1. Differential equations. 2. Topological dynamics.
3. Lifting theory. I. Sell, George R., 1937- joint author. II. Title. III. Series: American Mathematical Society. Memoirs ; no. 190.
QA3.A57 no. 190 [QA372] 515'.35 77-8941
ISBN 0-8218-2190-3

Table of Contents

Acknowledgements...iv

Abstract.. v

Chapter I. Introduction... 1

Chapter II. Statement of Main Results........................... 6

Chapter III. Applications to Ordinary Differential Equations....14

3.1 Definition of Basic Skew-Product Structure.......14

3.2 Compact Invariant Sets...........................15

3.3 The Distal Property..............................15

3.4 Stability Concepts...............................17

3.5 Equations Without Uniqueness.....................22

3.6 Scalar Equations.................................24

3.7 Triangularization of Linear Equations............28

Chapter IV. Other Applications..................................32

4.1 Retarded Differential Equations..................32

4.2 Kurzweil Equations...............................38

4.3 Volterra Equations...............................39

Chapter V. Proofs of Theorems 1,2 and 3.........................40

Chapter VI. An Example..49

Chapter VII. Flows on Fibre Bundles.............................52

Appendix A. Uniform Spaces......................................57

Appendix B. Distal Flows and the Enveloping Semigroups.......59

References..65

Abstract

Nonautonomous ordinary differential equations (ODE) $x' = f(x,t)$ are studied in the setting of the skew-product flow

$$\pi(x,f,t) = (\varphi(x,f,t), \sigma(f,t))$$

on a product space $X \times Y$, where X is an appropriate space of initial conditions and Y is a suitable space of functions $f = f(x,t)$. Sufficient conditions are given in order that certain dynamical structures on Y (e.g. almost periodicity) can be lifted to a compact π-invariant set $M \subseteq X \times Y$. In the ODE case, a lifting of the almost periodicity from Y to M is tantamount to proving the existence of an almost periodic solution. Our conditions for lifting are general enough to include most of the classical existence theorems for almost periodic solutions as special cases.

Our theory includes the study of skew-product semi-flows (i.e. flows defined only for $t \geq 0$), and consequently our results can be applied to the study of nonautonomous retarded functional differential equations (FDE). For example, if M is the ω-limit set of a bounded uniformly stable solution Ψ of an FDE $x'(t) = f(x_t,t)$ with almost periodic coefficients, then M is a distal minimal set in $X \times Y$ and the semi-flows π admits a unique extension to a flow $\hat{\pi}$ (defined for all $t \in R$) on M. If, in addition, Ψ is uniformly asymptotically stable, then M is a covering space of Y and the FDE $x'(t) = f(x_t,t)$ has an almost periodic solution.

The paper includes several illustrative examples.

Received by the Editors August 10, 1976.

LIFTING PROPERTIES IN SKEW-PRODUCT FLOWS WITH APPLICATIONS TO DIFFERENTIAL EQUATIONS

Chapter I

Introduction

In this paper we wish to study the lifting properties of skew-product flows. These properties are particulary important in the study of non-autonomous ordinary and functional differential equations. For example, the question of the existence of a periodic solution for a differential equation with periodic coefficients is the same as asking whether the periodic structure of the coefficient space can be lifted to a periodic structure in the solution space.

In order to reformulate the lifting problem more precisely we let W denote a topological Hausdorff space and let T denote the real numbers R, or the integers Z. Recall that a mapping $\pi: W\times T \to W$ is a flow if π is continuous, $\pi(w, 0) = w$ and $\pi(\pi(w,s), t) = \pi(w,s+t)$ for all $w \in W$ and $s,t \in T$. If $T = R$ one says that π is a continuous flow, and if $T = Z$, π is referred to as a discrete flow. If W is a product space $W = X\times Y$, then a flow π is said to be a skew-product flow if π has the form $\pi = (\varphi,\sigma)$, or

$$\pi(x,y,t) = (\varphi(x,y,t), \sigma(y,t))$$

where $\sigma: Y\times T \to Y$ is itself a flow on Y.

The study of discrete skew-product flows originated in ergodic theory, see especially Anzai [2], Furstenberg [17] and Parry and Walters [33]. Continuous skew-product flows arise in the study of ordinary differential equations and provide an appropriate geometric setting for nonautonomous equations, see Miller [26], Sell [42,43] and Sacker and Sell [35,37]. In the latter case, $X = R^n$ (or C^n), Y is a function space of functions $y = f$, where $f: R^n \times R \to R^n$ is Lipschitz-continuous, $\sigma(y,\tau)$ is the time - translation $\sigma(y,\tau) = f_\tau$ (where $f_\tau(x,t) = f(x,\tau+t)$), and finally $\varphi(x,y,\tau)$ represents the solution of the initial value problem

$$x' = f(x,t),\ x(0) = x$$

at time τ. We shall study this example in more detail in Chapter III. Other examples of skew-product flows are described in Sell [43; Chap. 3,11].

Let $\pi = (\varphi,\sigma)$ be a skew-product flow on $X \times Y$. We are interested in determining which properties of Y can be lifted to invariant sets M in $X \times Y$. This lifting problem is motivated by ordinary differential equations and is, for example, the basic problem one encounters when trying to determine whether there exist almost periodic solutions of almost periodic differential equations. More precisely, let us assume for the moment that Y is an almost periodic minimal set. (In the ordinary differential equations language, this means that Y is the hull of f, $Y = H(f) = Cl\{f_\tau : \tau \in R\}$, where $f(x,t)$ is Bohr almost periodic in t uniformly for x in compact subsets of $R^n = X$, cf. [26,43].) Next let us assume that there is a compact invariant set M in $X \times Y$. (This corresponds to assuming that the given differential equation $x' = f(x,t)$ has a solution which remains in a bounded subset of X for all $t \geq 0$. The invariant set M is then the ω-limit set of the corresponding motion, cf. [42,43].) Then we ask whether the flow on M is almost periodic, i.e. does the almost periodic structure on Y lift to M. (This corresponds to asking whether the given equation has an almost periodic solution, cf. [43].)

Probably the first results on the existence of almost periodic solutions for differential equations with almost periodic coefficients appeared in the book of Favard [13], who studied linear inhomogeneous equations. More recently there is the work of Amerio [1], who generalized some of Favard's results to nonlinear equations by introducing the concept of separatedness of solutions. Further extensions have been made by Seifert [40], who used the notion of global asymptotic stability to verify the separatedness condition, and Fink [15,16], who refined the separatedness condition and showed that it is related to the concept of uniform stability. By using somewhat different techniques, including various concepts of local stability, Miller [26] and Yoshizawa [49] have also studied this same basic question. On the surface the theories developed by Favard, Amerio, Seifert and Fink appear to be substantially different from those theories developed by Miller and Yoshizawa. But, as we shall see in Chapter III, both theories are consequences of the same general principle arising in the study of skew-product flows.

Our approach in this paper is then an abstraction of the theories developed for almost periodic ordinary differential equations. However our interest is not limited to almost periodic structures, but includes other structures such as minimality and distality. This approach will not only lead to interesting questions about the structure of skew-product flows, but will also lead to better insight into the behavior of solutions of non-autonomous ordinary and functional differential equations.

The general principle referred to above takes the form of three theorems. Theorem 1 is a general structure theorem for finite extensions of minimal transformation groups. Theorem 2 gives sufficient conditions under which an almost periodic structure on the base space Y can be lifted to a compact invariant set M in $X \times Y$. Theorem 3 is a reformulation of the above results in a context that is suitable for application to functional differential equations with almost periodic coefficients. The precise statements of these theorems will be given Chapter II, and the proofs will be deferred until Chapter V. These three theorems represent a generalization of our earlier work, [36].

In Chapter III we shall discuss various applications of our theory to ordinary differential equations. There we will show that the theory of Favard, Amerio, Seifert and Fink, as well as the theory of Miller and Yoshizawa are corollaries of Theorems 1 and 2. We shall also investigate in that section the theory of bounded solutions of scalar equations with almost periodic coefficients and show that to a limited extent one can derive an extension of the Poincaré- Bendixson theory for this case. Another interesting application is a global triangularization, by means of an almost periodic change of variables, of a second-order linear equation $x' = A(t)x$ with almost periodic coefficients.

In Chapter IV we shall discuss applications of our theory to time-varying functional differential equations. One of the results we prove is that if a retarded differential equation with almost periodic coefficients admits a positively compact uniformly stable solution, then the ω-limit set M of the corresponding motion is nonempty compact and minimal and, in addition, the semi-flow on M extends to a two-sided distal flow. As mentioned above, the proofs of our main theorems will be presented in Chapter V. In Chapter VI we shall discuss an example which shows that our theorems are best possible. Some of our results can be easily extended to fibre-preserving flows on a fibre bundle. This extension is discussed briefly in Chapter VII.

Since we shall be applying our results to general ordinary differential equations, including equations whose coefficient space is uniform but not metrizable, we formulate this theory in the context of uniform spaces. A brief review of uniform spaces is presented in Appendix A. Appendix B is concerned with some important properties of distal flows.

Finally we remark that the results which do not specifically depend on the special (natural) ordering of the group T can all be carried out, as in [36], for the case in which T belongs to the class $\mathfrak{J}$ of topological groups T having the property that there exists a compact subset $K \subseteq T$ such that T is generated by any open neighborhood of K. The results we are referring to are Theorem 1 (the equivalence of (A), (C) and (D)) and Theorem 2. See [25] where it is shown that the class $\mathfrak{J}$ is maximal with respect to proving certain dynamical properties.

We would like to express our appreciation to Professors R. Ellis, H. Keynes and L. Shapiro for their helpful comments and suggestions.

Chapter II

Statement of Main Results.

Because of our desire to apply this theory to the study of solutions of ordinary differential equations, it is appropriate to use the more general concept of a local flow, which permits one then to apply the theory of flows to differential equations with solutions that fail to be defined for all time t, for example, the equation $x' = x^2$.

Once again we let W be a topological Hausdorff space and let T denote the real numbers R, or the integers Z. For each $w \in W$ we let $I_w = (a_w, b_w)$ be an open interval in T with $0 \in I_w$, i.e. $a_w < 0 < b_w$. Let

$$D = \{(w,t) \in W \times T : t \in I_w\} .$$

For any transformation $\pi: D \to W$ we define the six sets

$$\gamma(w) = \{\pi(w,t): t \in I_w\}$$

$$\gamma^+(w) = \{\pi(w,t): t \in I_w \text{ and } t \geq 0\}$$

$$\gamma^-(w) = \{\pi(w,t): t \in I_w \text{ and } t \leq 0\}$$

$$H(w) = \mathrm{Cl}\ \gamma(w),\ H^+(w) = \mathrm{Cl}\ \gamma^+(w),\ H^-(w) = \mathrm{Cl}\ \gamma^-(w) ,$$

where Cl denotes the closure operation. Now given the family of intervals I_w and the set D, then a transformation $\pi: D \to W$ is said to be a local flow, or simply a flow, on W if

1) π is continuous,
2) $\pi(w,0) = w$ for all $w \in W$,
3) whenever $t \in I_w$ and $s \in I_{\pi(w,t)}$, then $(t+s) \in I_w$ and $\pi(\pi(w,t),s) = \pi(w,t+s)$,
4) the intervals I_w are maximal in the sense that either $I_w = T$,

or $H^+(w)$ is not compact when $b_w < +\infty$, and $H^-(w)$ is not compact when $a_w > -\infty$,

5) the intervals I_w are lower semi-continuous in w, i.e., if $w_n \to w$ in W then $I_w \subseteq \liminf I_{w_n}$.

In the event that $I_w = T$ for all $w \in W$, i.e., if $D = W \times T$, then π is referred to as a global flow. Actually the term "local flow", while widely used, may be a misnomer. One can show that if $\pi: D \to W$ is a flow on W, then π has no proper extension $\hat{\pi}$ to a similar domain $\hat{D}$ in $W \times T$, so that $\hat{\pi}: \hat{D} \to W$ is also a flow.

Let π be a flow on W. The function of t, $\pi(w,t)$ is referred to as the motion through w. The point set $\gamma(w)$ defined above is referred to as the trajectory through w and $H(w)$ is called the hull of w. A set $A \subseteq W$ is said to be invariant if $\gamma(w) \subseteq A$ whenever $w \in A$. A is positively invariant if $\gamma^+(w) \subseteq A$ whenever $w \in A$. It is easily seen that if A is invariant, then the closure $\mathrm{Cl}\, A$ is also an invariant set. Notice that if A is a compact invariant set in W, then (as a consequence of property (4)) for all $w \in A$ one has $I_w = T$, i.e., the motion $\pi(w,t)$ is defined for all $t \in T$. If a motion $\pi(w,t)$ is defined for all $t \geq 0$, then the ω-limit set is defined by

$$\Omega(w) = \cap_{\tau \geq 0} H^+(\pi(w,\tau)) .$$

If W is a product space $W = X \times T$, then a flow π is said to be a skew-product flow if π has the form $\pi = (\varphi, \sigma)$, or

$$\pi(x,y,t) = (\varphi(x,y,t), \sigma(y,t)) ,$$

where σ is independent of x. This means that σ admits an extension which is a flow on Y. In this paper we shall assume that $\pi = (\varphi, \sigma)$ is a given skew-product flow on $X \times Y$, where X is metric space with a metric d and Y is a compact Hausdorff space, with the induced uniform topology, cf. Appendix A. Since Y is compact that means that the flow σ is a global flow on Y. Also we should note that nonmetrizable uniform spaces do arise in the study of differential equations when one introduces various weak topologies on appropriate function spaces. (See, for example, the G_p - spaces described in [28, pp 19-20].) The assumption that X be metrizable seems to be essential in the proof of our main results. Fortunately this restriction is natural in the context of differential equations, since X is usually a subset of either some Euclidean space, or some manifold, or perhaps, some Banach space.

Let $\pi = (\varphi, \sigma)$ be a skew-product flow on $X \times Y$ and let $M \subseteq X \times Y$ be a compact π-invariant set. Let $p: X \times Y \to Y$ be the natural projection and define

$$\mu(y) = \operatorname{card}(p^{-1}(y) \cap M), \qquad y \in Y .$$

If Y is a compact minimal set, then M covers Y, i.e., $\mu(y) \geq 1$ for all $y \in Y$. We shall say that $p|M$, the restriction of p to M, is of <u>distal type</u> (and that the invariant set M has the <u>fibre - distal property</u>) if for any two points (x_1, y) and (x_2, y) in $p^{-1}(y) \cap M$ with $x_1 \neq x_2$, there is an $\alpha = \alpha(x_1, x_2, y) > 0$ such that

$$\text{(2.1)} \qquad d(\varphi(x_1, y, t), \varphi(x_2, y, t)) \geq \alpha , \qquad (\text{for all } t \in T) .$$

If (2.1) holds only for $t \geq 0$ (or $t \leq 0$), then $p|M$ is said to be of <u>positive</u> (or <u>negative</u>) <u>distal type</u>. We can now state our first result, the Structure Theorem.

Theorem 1. *Let* $\pi = (\varphi, \sigma)$ *be a skew-product flow on* $X \times Y$, *where* Y *is a compact minimal set in the flow* σ *and* X *is metrizable. Let* $M \subseteq X \times Y$ *be a compact* π*-invariant subset. Then the following statements are equivalent:*

(A) $p|M$ *is of distal type and there is a* $y_o \in Y$ *such that* $\mu(y_o) = N$ *is finite.*

(B) $p|M$ *is of positive (or negative) distal type and there is a* $y_o \in Y$ *such that* $\mu(y_o) = N$ *is finite.*

(C) $\mu(y) \equiv N < \infty$ *for all* $y \in Y$.

(D) M *is an* N*-fold covering space of* Y *with covering map* $p|M$.

If the above conditions are satisfied, then M *can be expressed as a finite union of minimal sets* $M_1, \ldots, M_k$, *where* M_i *is an* n_i*-fold covering of* Y *and* $n_1 + \ldots + n_k = N$.

As we shall see, Theorem 1 turns out to be a useful fact to employ when determining which dynamical properties lift from Y to M. In order to formulate an application along these lines we need to recall two concepts: Distality and almost periodicity.

Let M be a compact invariant set for a flow π on a Hausdorff space W and let $\mathcal{P} = \{\rho\}$ denote a family of pseudo-metrics which generate the topology on M. (It is not necessary, for this purpose, that W be a uniform space, although in practice, it would be. The important point here is that, regardless of the topology on W, the induced topology on the compact set M is uniform, cf. Appendix A.) The flow π is said to be *distal* on M if for any two distinct points $w_1, w_2 \in M$, there is a $\rho \in \mathcal{P}$ and an $\alpha > 0$ such that

$$\rho(\pi(w_1,t), \pi(w_2,t)) \geq \alpha, \qquad \text{(for all } t \in T). \tag{2.2}$$

Another way of saying this is that the flow π is distal on M if and only if

$$\lim \pi(w_1,t_n) = \lim \pi(w_2,t_n) \Rightarrow w_1 = w_2 \quad ,$$

where $\{t_n\}$ is any sequence is T for which the above limits exist. (Here, and in what follows, "sequence" means generalized sequence or net whenever appropriate.) By replacing $t \in T$ with $t \geq 0$ or $t \leq 0$ in (2.2) one has the corresponding notions of <u>positive</u> or <u>negative distality</u>. The flow π is <u>almost periodic</u> (<u>equi-continuous</u>) on M if for every $\rho \in \mathcal{P}$ and $\nu > 0$ there is a $\hat{\rho} \in \mathcal{P}$ and a $\delta > 0$ such that

$$(2.3) \qquad \rho(\pi(w_1,t), \pi(w_2,t)) < \nu , \qquad (\text{for all } t \in T) ,$$

whenever $w_1,w_2 \in M$ and $\hat{\rho}(w_1,w_2) < \delta$.

Clearly almost periodicity implies distality. The converse is not true in general, cf. [11,22]. However the following concept is equivalent to almost periodicity: The flow π is said to be <u>uniformly distal</u> on M if for every $\rho \in \mathcal{P}$ and $\nu > 0$ there is a $\hat{\rho} \in \mathcal{P}$ and a $\delta > 0$ such that

$$(2.4) \qquad \hat{\rho}(\pi(w_1,t) , \pi(w_2,t)) \geq \delta \qquad (\text{for all } t \in T) ,$$

whenever $w_1,w_2 \in M$ and $\rho(w_1,w_2) \geq \nu$.

Before stating the next theorem we recall that if Y is a uniform space and X is a metric space with metric d , then $X \times Y$ is a uniform space. Indeed, if ρ is a pseudo-metric on Y , then $d \cdot \rho$ is a pseudo-metric on $X \times Y$ where

$$d \cdot \rho \ (w,\hat{w}) = \max\{ d(x,\hat{x}), \rho(y,\hat{y}) \} \quad ,$$

$w = (x,y)$ and $\hat{w} = (\hat{x},\hat{y})$. Moreover, if $\mathcal{P} = \{\rho\}$ is a family of pseudo-metrics which generate the uniform topology on Y, then $\mathfrak{D} = \{d \cdot \rho : \rho \in \mathcal{P}\}$ generates the product topology on $X \times Y$.

The following result, which is an application of Theorem 1, is an example of a "lifting theorem."

Theorem 2. Assume that the hypotheses of Theorem 1 are satisfied and that any one (and hence all) of the conditions (A) through (D) is satisfied. Then the following statements are valid:

(E) If the flow σ on Y is distal then then the flow π on M is distal.

(F) If the flow σ on Y is almost periodic, then the flow π on M is almost periodic.

Remarks 1. In the case where both X and Y are metrizable, Theorems 1 and 2 follow from general structure theorems for finite extensions of minimal transformation groups, cf. [36].

2. One says that a compact invariant set M laminates if M is the union of minimal sets. One of the assertions of Theorem 1 is that the given compact invariant set M laminates. This particular fact has been established under less restrictive conditions than those used in Theorem 1. For example, Ellis [10] has shown that if the flow on a compact invariant set M is distal, then M laminates. More generally Auslander [5] has shown that if M is a compact invariant set in a skew-product flow on $X \times Y$, where Y is a compact minimal set, then M laminates whenever $p|M$ is of distal type.

3. The equivalence between our definition of almost periodicity and the concepts of Bohr and Bochner is discussed in Sell [45] and Sacker and Sell [39]. Also see Fink [16] and Nemytskii and Stepanov [32]. It should be noted that some authors, see [18] for example, refer to our notion of

almost periodicity as "unifrom almost periodicity".

4. Conclusions (E) and (F) in Theorem 2 can be made stronger by stating that "the flow σ on Y is almost periodic (or distal) if and only if the flow π on M is almost periodic (or distal)", cf. [12,36].

Theorems 1 and 2, as we shall see in the next chapter, can be applied directly to ordinary differential equations. Functional differential equations (or differential-delay equations) are a different matter. The problem here is that the solutions of such equations prefer positive time, i.e., the solutions need not exist for time $t < 0$. What is required in that case is the notion of a <u>semi-flow</u>, which differs from a flow in that the function $\pi(w,t)$ need not be defined for $t < 0$. For a precise definition of a semi-flow one replaces the interval $I_w = (a_w, b_w)$ with a half-open interval $[0, b_w)$ and then makes the obvious modifications in the formulation given above. We shall omit the details.

In order to emphasize the difference between a flow and a semi-flow, we shall refer to the former as a <u>two-sided flow</u> and the latter as a <u>one-sided flow</u>. Simple examples will show that in some situations a one-sided flow can be extended to a two-sided flow, however this extension is not automatic, cf. [7].

We now have the following result.

<u>Theorem</u> 3. <u>Let</u> $\pi = (\varphi, \sigma)$ <u>be a skew-product semi-flow on</u> $X \times Y$, <u>where</u> X <u>is metrizable and</u> Y <u>is compact</u>. <u>Assume that</u> σ <u>can be extended to a two-sided flow on</u> Y <u>and</u> Y <u>is minimal in this flow</u>. <u>Let</u> $M \subseteq X \times Y$ <u>be a compact positively invariant subset for the semi-flow</u> π. <u>Then the following statements are equivalent</u>:

(A) <u>The restriction of</u> π <u>to</u> M <u>can be extended to a two-sided flow with the following properties</u>: $p|M$ <u>is of distal type and there is a</u>

$y_o \in Y$ *such that* $\mu(y_o) = N$ *is finite*.

(B) $p|M$ *is of positive distal type and there is a* $y_o \in Y$ *such that* $\mu(y_o) = N$ *is finite*.

(C) $\mu(y) \equiv N < \infty$ *for all* $y \in Y$.

(D) M *is an* N-*fold covering space of* Y *with covering map* $p|M$.

If the above conditions hold then the remaining conclusions of Theorem 1 *and statements* (E) *and* (F) *in Theorem* 2 *are valid*.

Chapter III

Applications to Ordinary Differential Equations

3.1 Definition of Basic Skew-Product Structure. Let $\mathfrak{F}$ denote a collection of functions $f: X \times R \to R^n$ where X is an open set in R^n. We assume that $\mathfrak{F}$ is translation-invariant, i.e. $f_\tau \in \mathfrak{F}$ whenever $f \in \mathfrak{F}$ and $\tau \in R$, where $f_\tau(x,t) = f(x,\tau+t)$. In addition we assume that $\mathfrak{F}$ has a topology so that the mapping

$$\sigma(f,\tau) = f_\tau \tag{3.1}$$

is a continuous mapping of $\mathfrak{F} \times R$ into $\mathfrak{F}$, cf. [3,29] for example. The space $\mathfrak{F}$ will take the role of the base space Y in our construction of the skew-product flow.

In addition to the above we now assume that $\mathfrak{F}$ satisfies the following two properties:

(i) The Carathéodory Property: For every $f \in \mathfrak{F}$ and every $x_o \in X$ there is a unique noncontinuable solution $\varphi(x_o,f,t)$ of the initial value problem

$$x' = f(x,t),\ x(0) = x_o$$

defined on some open interval $I_{(x_o,f)}$ containing $t_o = 0$.

(ii) The Kamke Property: If $\{\varphi_n\}$ is a sequence of noncontinuable solutions of

$$x' = f_n(x,t), \varphi_n(t_o) = x_n,$$

where $f_n \to f$ in $\mathfrak{F}$ and $\varphi_n(t_o) \to x_o$ in X, then one has $\varphi_n(t) \to \varphi(t)$ uniformly for t in compact subsets of the interval of definition of φ, where φ is the noncontinuable solution of the initial value problem

$$x' = f(x,t), \; x(t_o) = x_o \; .$$

Under these conditions a skew-product flow is defined on $X \times \mathfrak{F}$ by

$$(3.2) \qquad \pi(x,f,\tau) = (\varphi(x,f,\tau), \; \sigma(f,\tau)) \; ,$$

cf. Sell [42,43] .

3.2 Compact Invariant Sets. The construction of the skew-product flow above is very general. For our purposes we shall want to restrict the family $\mathfrak{F}$ to be compact in the given topology. For example, assume that $\mathfrak{F}_o$ is a translation-invariant family of continuous functions with the topology of uniform convergence on compact sets in $X \times R$. If there is a constant k such that

$$|f(x_1,t_1) - f(x_2,t_2)| \leq k[\; |x_1 - x_2| + |t_1 - t_2| \;]$$

for all $f \in \mathfrak{F}_o$ and all $x_1,x_2 \in X$, $t_1,t_2 \in R$, then the closure $\mathfrak{F} = \mathrm{Cl}\; \mathfrak{F}_o$ is compact and satisfies both the Carathéodory and the Kamke properties. (cf. [3,43,48] for other examples.)

In the case that $\mathfrak{F}$ is compact it is easy to show that there is a compact invariant set $M \subseteq X \times \mathfrak{F}$ if and only if there is $f \in \mathfrak{F}$ and $x \in X$ such that the solution $\varphi(x,f,t)$ remains in a compact subset of X for all $t \geq 0$. Indeed, the set M can be chosen to be the ω-limit set of the motion through $(x,f) \in X \times \mathfrak{F}$, i.e., $M = \Omega(x,f)$ cf. [42].

3.3 The Distal Property. Now let M be a compact invariant set in $X \times \mathfrak{F}$, where we assume that $\mathfrak{F}$ is compact. Let $p\colon X \times \mathfrak{F} \to \mathfrak{F}$ be the natural projection. Recall that M has the fibre-distal property if given two points (x_1,f) , $(x_2,f) \in M$ with $x_1 \neq x_2$ then there is an $\alpha = \alpha(x_1,x_2,f) > 0$ such that

$$d(\varphi(x_1,f,t),\ \varphi(x_2,f,t)) \geq \alpha$$

for all $t \in R$.

In order to compare this concept with various separatedness conditions used by other authors, it is convenient to reformulate the distal property in a slighty more restrictive manner: Let K be a compact set in X . We say that the solutions of $x' = f(x,t)$ (where f is a fixed element of $\mathfrak{F}$) are weakly separated in K if given $x_1, x_2 \in K$ satisfying

(i) $x_1 \neq x_2$, and

(ii) $\varphi(x_1,f,t)$ and $\varphi(x_2,f,t)$ remain in K for all t in R ,

then there is an $\alpha = \alpha(x_1,x_2,f) > 0$ such that

$$d(\varphi(x_1,f,t),\ \varphi(x_2,f,t)) \geq \alpha\ , \qquad \text{(for all } t \text{ in } R). \tag{3.3}$$

Now if it happens that for every $f \in \mathfrak{F}$ the solutions of $x' = f(x,t)$ are weakly separated in K , then any compact π-invariant set M contained in $K \times \mathfrak{F}$ has the fibre-distal property.

Amerio [1] defines the solutions of $x' = f(x,t)$ to be separated in K if there is a constant $\alpha = \alpha(f) > 0$ such that if $x_1, x_2 \in K$ and satisfy (i) and (ii) above, then (3.3) holds for all $t \in R$. Fink [15] introduces an apparently weaker concept by defining the solutions of $x' = f(x,t)$ to be semi-separated in K if given any solution $\varphi(x_1,f,t)$ that remains in K for all t , then there is an $\alpha = \alpha(x_1,f) > 0$ such that if $\varphi(x_2,f,t)$ is any other solution (with $x_1 \neq x_2$) that remains in K for all t , then (3.3) holds for all t in R . (Actually Fink only requires (3.3) to hold for all $t \geq 0$. This variation, in the context of this paper, is more apparent than real, cf. Remark 5 (p.21) and Appendix B.)

The following implications are direct consequences of the definitions:

Separated => Semi-separated => Weakly separated.

As a matter of fact, if the solutions of $x' = f(x,t)$ are semi-separated in a compact set K, then there are at most finitely many solutions that remain in K for all t in R. From this it follows that the concepts of "separatedness" and "semi-separatedness" are the same. Putting it another way, the assumption of separated solutions in the sense of Amerio-Fink implies both the distal-type property and the finiteness property $\mu(y) = N < \infty$ used in Theorem 1. (Weakly separated, on the other hand, does not imply the finiteness property.) More precisely, if for every $f \in \mathfrak{F}$ the solutions of $x' = f(x,t)$ are separated in a compact set K, and if M is any compact π-invariant set in $K \times \mathfrak{F}$, then M has the fibre-distal property and

$$\mu(f) = \text{card}\,(p^{-1}(f) \cap M) < \infty$$

for all $f \in \mathfrak{F}$. Based on these facts they then argued the existence of an almost periodic solution. One would hope that by segregating the distal property and the finiteness property, the arguments using Theorems 1,2 and 3 can be simplified; and as we shall see, this hope is justified.

3.4 Stability Concepts. Several authors, cf. [14, 15, 31], have observed connections between various stability concepts and the separatedness of solutions. This connection becomes easier to understand (in our opinion) if one uses properties of distal flows together with Theorems 1 and 2. Our starting point is that certain uniform stability properties are inherited by the limiting equations.

Throughout the remainder of this section we will assume that $\pi = (\varphi,\sigma)$ is a skew-product flow on $X \times \mathfrak{F}$, as described above. In particular we assume that $\mathfrak{F}$ satisfies the Carathéodory and Kamke properties.

Recall that a solution $\varphi(x,f,t)$ of $x' = f(x,t)$ is said to be <u>uniformly stable</u> if $\varphi(x,f,t)$ is defined for all $t \geq 0$ and for every $\nu > 0$ there is a $\delta = \delta(\nu) > 0$ such that

$$|\varphi(x,f,\tau+t) - \varphi(\hat{x},f,\tau+t)| \leq \nu \qquad \text{(for all } t \geq 0) ,$$

whenever $\tau \geq 0$ and $|\varphi(x,f,\tau) - \varphi(\hat{x},f,\tau)| \leq \delta$. (The function $\delta(\nu)$ is referred to as the <u>modulus of uniform stability</u> for $\varphi(x,f,t)$.) The solution $\varphi(x,f,t)$ is said to be <u>uniformly asymptotically stable</u> if it is uniformly stable and there is a $\delta_o > 0$, and for every $\nu > 0$ there is a $t_o > 0$ such that

$$|\varphi(x,f,\tau+t) - \varphi(\hat{x},f,\tau+t)| \leq \nu , \qquad \text{(for } t \geq t_o) ,$$

whenever $\tau \geq 0$ and $|\varphi(x,f,\tau) - \varphi(\hat{x},f,\tau)| \leq \delta_o$.

A solution $\varphi(x,f,t)$ of $x' = f(x,t)$ is said to be <u>compact</u> (or <u>positively compact</u>) if $\varphi(x,f,t)$ remains in a compact subset of X for all $t \in R$ (or $t \geq 0$). The following result is a basic stability theorem, cf. [42,43]:

<u>Theorem</u> 4. <u>Let</u> $\varphi(x,f,t)$ <u>be positively compact solution of</u> $x' = f(x,t)$ <u>and assume that</u> $\varphi(x,f,t)$ <u>is uniformly stable. Then for every</u> (x^*,f^*) <u>in</u> $\Omega(x,f)$ <u>the solution</u> $\varphi(x^*,f^*)$ <u>is defined for all</u> $t \in R$; <u>furthermore this solution is compact and uniformly stable. Moreover,</u> $\varphi(x^*,f^*,t)$ <u>has the same modulus of uniform stability as</u> $\varphi(x,f,t)$. <u>If, in addition,</u> $\varphi(x,f,t)$ <u>is uniformly asymptotically stable, then for every</u> (x^*,f^*) <u>in</u> $\Omega(x,f)$,

the solution $\varphi(x^*,f^*,t)$ *is uniformly asymptotically stable.*

In the case that $\mathfrak{F}$ is an almost periodic minimal set, one can say more.

Theorem 5. *Let* $\mathfrak{F}$ *be an almost periodic minimal set and assume that there exists a positively compact solution* $\varphi(x,f,t)$ *that is uniformly stable. Then the* ω*-limit set* $\Omega(x,f)$ *is a nonempty compact minimal set in* $X\times\mathfrak{F}$ *and the flow* π *is distal on* $\Omega(x,f)$.

Proof: Since the motion $\pi(x,f,t)$ remains in a compact set in $X\times\mathfrak{F}$ for all $t\geq 0$, it follows that $\Omega(x,f)$ contains a minimal set M . We will show later that $\Omega(x,f)=M$. But let us first show that the flow π is distal on $\Omega(x,f)$. For this purpose, it will suffice to show that $\Omega(x,f)$ has the fibre-distal property, since the flow on the base space $\mathfrak{F}$ is almost periodic and *a fortiori* distal. In fact it will suffice to show that if (x_1,g), $(x_2,g)\in\Omega(x,f)$ with $x_1\neq x_2$ then there is an $\alpha>0$ such that

$$(3.4)\qquad |\varphi(x_1,g,t)-\varphi(x_2,g,t)|\geq\alpha\ ,\qquad (\text{for all}\quad t\leq 0)\ .$$

Indeed (3.4) implies that the flow on $\Omega(x,f)$ is negatively distal and (as is shown in Appendix B) this implies that the flow π is distal on $\Omega(x,f)$.

In order to prove (3.4) we use the uniform stability. Let $\delta(\nu)$ be the modulus of uniform stability for $\varphi(x,f,t)$. Then by Theorem 4, the same $\delta(\nu)$ is the modulus of uniform stability for every solution $\varphi(x^*,f^*,t)$ where $(x^*,f^*)\in\Omega(x,f)$. Define ν_0 by $2\nu_0=|x_1-x_2|>0$. Set $\delta_0=\delta(\nu_0)$. If (3.4) does not hold, then there is a $\tau<0$ such that $|\varphi(x_1,g,\tau)-\varphi(x_2,g,\tau)|\leq\delta_0$. Since $(\varphi(x_1,g,\tau),\ g_\tau)\in\Omega(x,f)$ and $\varphi(\varphi(x_1,g,\tau),g_\tau,t)=\varphi(x_1,g,\tau+t)$, one then has $|\varphi(x_1,g,\tau+t)-\varphi(x_2,g,\tau+t)|\leq\nu_0$ for all $t\geq 0$. In particular, at $t=-\tau$ we get the contradiction

$$2\nu_o = |x_1 - x_2| = |\varphi(x_1,g,\tau+t) - \varphi(x_2,g,\tau+t)| \leq \nu_o \quad .$$

Let us now show that $\Omega(x,f) = M$. For this purpose, fix $\nu > 0$ and choose $(x_1,f_1) \in M \subseteq \Omega(x,f)$. Since $\Omega(x,f)$ has the fibre-distal property, it follows that $p_1 = p|M$, the restriction of p to M, is of distal type. Consequently the mapping $p_1 \colon M \to \mathfrak{F}$ is open, cf. Appendix B. Choose $\eta = \frac{1}{2}\delta(\nu)$. Define the (large) open set $\mathcal{V} = \{(x,f) \in X \times \mathfrak{F} \colon |x-x_1| < \eta\}$ and set $\mathcal{V}_1 = \mathcal{V} \cap M$. Since $p_1 = p|M$ is open, the set $p_1(\mathcal{V}_1)$ is open in $\mathfrak{F}$. Choose U to be an open subset in $p_1(\mathcal{V}_1)$ which contains f_1. Finally define

$$V' = \{(x,f) \in X \times \mathfrak{F} \colon |x-x_1| < \eta,\ f \in U\}$$

Then V' is an open neighborhood in $X \times \mathfrak{F}$ of (x_1,f_1) having the following property: For each point $g \in U$ there exists a point $(\xi,g) \in M \cap V'$, and in particular, one has $|\xi-x_1| < \eta$.

Now since $(x_1,f_1) \in \Omega(x,f)$, there is a point $(\hat{x},\hat{f}) \in \gamma^+(x,f) \cap V'$. Thus $\hat{f} \in U$ and, from the above property, there is a point $(x_2,\hat{f}) \in M \cap V'$. Thus $\hat{x}$ and x_2 lie in the fibre over $\hat{f}$ and

$$|x_2 - \hat{x}| \leq |x_2 - x_1| + |x_1 - \hat{x}| < 2\eta = \delta(\nu) \quad .$$

Therefore

$$|\varphi(x_2,\hat{f},t) - \varphi(\hat{x},\hat{f},t)| \leq \nu$$

for all $t \geq 0$. Since $\pi(x_2,\hat{f},t) \in M$ for $t \geq 0$ and $\Omega(x,f) = \Omega(\hat{x},\hat{f})$, one sees that points of $\Omega(x,f)$ lie within ν of points of M (the distance being measured in each fibre). Since ν is arbitrary one has $\Omega(x,f) = M$.

QED.

We see then that uniform stability, in the case of differential equations with almost periodic coefficients, gives the distal property. The stronger condition of uniform asymptotic stability implies the finiteness property.

Theorem 6. Let $\mathfrak{F}$ be an almost periodic minimal set and assume that there exists a positively compact solution $\varphi(x,f,t)$ that is uniformly asymptotically stable. Then $\Omega(x,f)$ is an almost periodic minimal set and is an N-fold covering space of $\mathfrak{F}$, where N is finite. In particular for every $(x^*,f^*) \in \Omega(x,f)$ the solution $\varphi(x^*,f^*,t)$ is almost periodic.

Proof: It follows from Theorem 5 that $\Omega(x,f)$ has the fibre-distal property. Also it follows from Theorem 4 that for each $(x^*,f^*) \in \Omega(x,f)$, the solution $\varphi(x^*,f^*,t)$ is compact and uniformly asymptotically stable. Now if some fibre $p^{-1}(g) \cap \Omega(x,f)$ contains an infinite number of points, then it must contain an accumulation point, which we shall denote by (v,g). Since (v,g) is an accumulation point of $p^{-1}(g) \cap \Omega(x,f)$, there is a point $(u,g) \in \Omega(x,f)$ such that $u \neq v$ and $|u-v|$ is small. It follows from the uniform asymptotic stability that $|\varphi(u,g,t) - \varphi(v,g,t)| \to 0$ as $t \to +\infty$, which contradicts the fibre-distal property. Hence every fibre $p^{-1}(g) \cap \Omega(x,f)$ contains a finite number of points. It follows from Theorem 1 that $\Omega(x,f)$ is an N-fold cover of $\mathfrak{F}$. By Theorem 2, $\Omega(x,f)$ is the union of almost periodic minimal sets. Since $\Omega(x,f)$ is minimal (Theorem 5) the proof is now complete. QED.

Remarks 5. In the proof of Theorem 4 we used the fact that negative distality (distal for $t \leq 0$) implies two-sided distal (distal for all $t \in R$). This implication is discussed in Appendix B. The same implication, in the case of differential equations with almost periodic coefficients, shows that the positive semi-separatedness concept of Fink (i.e., (3.3) holds for all

$t \geq 0$) is equivalent to the two-sided semi-separatedness concept defined in Section 3.3(i.e., (3.3) holds for all $t \in R$.)

6. Theorem 6 was first proved by Miller [26] although his proof was substantially different. He used the fact that a positively compact solution $\varphi(x,f,t)$, that is uniformly asymptotically stable, is stable under persistent disturbances provided the equations in $\mathfrak{F}$ are locally Lipschitz-continuous in x with Lipschitz constants independent of t, cf. Malkin [24]. The latter type of stability, in our language, implies that for every $\nu > 0$ there is a $\delta > 0$ such that whenever $(u,g),(v,h) \in \Omega(x,f)$ with $|u-v| \leq \delta$ and $d(g,h) \leq \delta$ (the space $\mathfrak{F}$ is metrizable with a metric d in Miller's case,) then $|\varphi(u,g,t) - \varphi(v,h,t)| \leq \nu$ for all $t \geq 0$. But once one has this property, the battle is over; because of Markov's Theorem ([32,p. 390]) the motion $\pi(u,g,t)$ is almost periodic. Yoshizawa [50] also develops the theory of almost periodicity by using Malkin's stability theory and the theory of Lyapunov functions.

7. We see then that our Theorems 1 and 2, together with Theorem 5, represents the general dynamical principle which leads to the Miller-Yoshizawa theory, on the one hand, and the Favard-Amerio-Fink-Seifert theory, on the other.

8. In general the conclusion of Theorem 5 cannot be strengthened to conclude that $\Omega(x,f)$ is an almost periodic minimal set. The simplest example is the complex-valued equation $z' = i\,a(t)\,z$ where $a(t)$ is a real-valued almost periodic function with mean value zero and unbounded integral $A(t) = \int_0^t a(s)ds$. In this case the solution $\varphi(x,a,t) = e^{iA(t)}x$ is compact and uniformly stable but not almost periodic when $x \neq 0$, cf. [16, p. 106].

3.5 <u>Equations Without Uniqueness</u>. The general theory described in Theorems 1 and 2 is also applicable to ordinary differential equations lacking uniqueness. The basic idea here can perhaps be best explained for ordinary differential equations with continuous coefficients. (Let us note here that some special problems related to the one-sided non-uniqueness arising in functional differential equations will be discussed in the next chapter.)

Let $\mathfrak{F}$ denote a collection of continuous functions $f: R^n \times R \to R^n$ and assume that $\mathfrak{F}$ has the topology of uniform convergence on compact sets. Consequently $\sigma(f,\tau) = f_\tau$ is a flow on $\mathfrak{F}$.

Next let X denote the collection of all continuous functions φ with the following properties:

(i) Each $\varphi \in X$ is a continuous function from an open interval $I_\varphi = (a_\varphi, b_\varphi)$ (which may depend on φ) to R^n, and $0 \in I_\varphi$.

(ii) Each φ is noncontinuable, i.e., if $b_\varphi < +\infty$ then $|\varphi(t)| \to +\infty$ as $t \to b_\varphi^-$ and if $-\infty < a_\varphi$ then $|\varphi(t)| \to +\infty$ as $t \to a_\varphi^+$.

A topology is defined on X by saying that $\varphi_n \to \varphi$ as $n \to \infty$, if $I_\varphi \subseteq \liminf I_{\varphi_n}$ and $\varphi_n(t)$ converges to $\varphi(t)$ uniformly for t in compact sets in I_φ. This topology is metrizable. Let

$$D = \{(\varphi, f, \tau) \in X \times \mathfrak{F} \times R : \tau \in I_\varphi\}$$

It is now easy to verify that the mapping $\pi: D \to X \times \mathfrak{F}$ defined by

$$\pi(\varphi, f, \tau) = (\varphi_\tau, f_\tau),$$

where $\varphi_\tau(t) = \varphi(\tau + t)$ and $f_\tau(x,t) = f(x, \tau + t)$, is a (local) skew-product flow on $X \times \mathfrak{F}$. It is shown in [44] that the set $W = \{(\varphi, f): \varphi \in \mathcal{S}(f)\}$, where $\mathcal{S}(f)$ denotes the collection of all non-continuable solutions of $x' = f(x,t)$, is a π-invariant set. Therefore if $M \subseteq W$ and M is π-invariant, then Theorems 1 and 2 are applicable to M.

The Amerio-Fink theory of almost periodic solutions, which is based on the separatedness of solutions, can now be extended without any modification to differential equations without uniqueness. The Miller-Yoshizawa theory, which is based on local stability, is another matter. The difficulty in the latter case occurs in Theorem 4. This theorem can break down if the uniqueness of solutions is dropped, cf. Chapter VI. For equations lacking uniqueness one needs a different form of stability such as total stability or stability under persistent disturbances, cf. [26,49,50] and Remark 6.

3.6 Scalar Equations. In the case of scalar-valued differential equations, i.e., $x' = f(x,t)$ where $x \in R^1$, one can use the ordering of R^1 to derive some additional information about the structure of the induced skew-product flow. For this purpose we will need the following lemma, which we formulate for a skew-product flow.

Lemma 7. *Let $\pi = (\varphi,\sigma)$ be a continuous $(T=R)$ skew-product flow on $X \times Y$, where $X = R^1$. Let $x_1, x_2 \in X$ with $x_1 < x_2$. Then the following assertions are valid:*

(A) *$\varphi(x_1,y,t) < \varphi(x_2,y,t)$ for all $y \in Y$ and all $t \in I_{(x_1,y)} \cap I_{(x_2,y)}$, where $I_{(x,y)}$ denotes the interval of definition for the motion $\pi(x,y,t)$.*

(B) *If $x_i^* = \lim \varphi(x_i,y,t_n)$, $i=1,2$, for some sequence $\{t_n\}$ and some $y \in Y$, then $x_1^* \le x_2^*$.*

The proof is straightforward and we omit the details. Now by using this lemma one can prove the following:

Theorem 8. *Let $\pi = (\varphi,\sigma)$ be a skew-product flow on $X \times \mathfrak{F}$, as described above. Assume that $X = R^1$ and that $\mathfrak{F}$ is a compact almost periodic minimal set. Assume further that there is a positively compact motion $\varphi(x,f,t)$ that is uniformly stable. Then the ω-limit set $\Omega(x,f)$ is a 1-cover of the base $\mathfrak{F}$. In particular, $\Omega(x,f)$ is an compact* almost periodic minimal set.

Proof: It follows from Theorem 5 that $\Omega(x,f)$ is a compact minimal set and that the flow π is distal on $\Omega(x,f)$. Because of Theorems 1 and 2 it will suffice to show that each fibre $p^{-1}(g) \cap \Omega(x,f)$, $g \in \mathfrak{F}$, contains a single point. We shall do this by contradiction. Assume there is a fibre $p^{-1}(g) \cap \Omega(x,f)$ with at least two points. Since the fibre is compact, there is a unique minimum and maximum, x_1 and x_2. That is, there exist $x_1, x_2 \in X = R^1$ such that $x_1 < x_2$, $(x_i, g) \in \Omega(x,f)$, $i = 1,2$, and for all $(x^*, g) \in \Omega(x,f)$ one has $x_1 \leq x^* \leq x_2$. Since $\Omega(x,f)$ is minimal there is a sequence $\tau_n \to +\infty$ such that

$$\pi(x_1, g, \tau_n) \to (x_2, g) .$$

In particular $x_2 = \lim \varphi(x_1, g, \tau_n)$. One can then choose a subsequence of $\{\tau_n\}$, which we again denote by $\{\tau_n\}$, so that $\lim \varphi(x_2, g, \tau_n)$ exists. Call this limit x_3. One then has $(x_3, g) \in \Omega(x,f)$, since $\Omega(x,f)$ is closed and invariant. Furthermore one has $x_2 \leq x_3$ by Lemma 7. However one cannot have $x_2 < x_3$, since x_2 is the maximum. Hence one has $x_2 = x_3$, which contradicts the distality of the flow on $\Omega(x,f)$. QED.

The last theorem is of interest because, in the scalar-valued case, one can drop the asymptotic stability assumption in Theorem 6 and still conclude that the ω-limit set is an almost periodic minimal set. The uniform stability assumption, however, cannot in general be dropped, even in the case of linear scalar-valued equations. The following example, which was suggested by A.M. Fink, illustrates the difficulty.

Example. Consider the linear equation

$$x' = -f(t)x \tag{3.5}$$

where $f(t) = \sum_{k=1}^{\infty} f_k(t)$ and

$$f_k(t) = 2^{-k} \pi \sin (2^{-k} \pi t) \quad .$$

Since f is the uniform limit of periodic functions, it is almost periodic. Furthermore one has

$$\int_0^t f(s)ds = \Sigma_{k=1}^{\infty} \int_0^t f_k(s)ds$$

for all $t \in R$. The last equality then implies that

$$\int_0^t f(s)ds \geq 0 \; , \qquad (\text{for all } t \geq 0) \; ,$$

and

$$A_m(t) \leq \int_0^t f(s)ds \leq A_m(t) + 2\pi \; , \qquad (\text{for } 0 \leq t \leq 2^{m+1}),$$

where $A_m(t) = \Sigma_{k=1}^{m} \int_0^t f_k(s)ds$, $m=1,2,\ldots.$ In addition, one has

$$(3.6) \qquad \int_0^t f(s)ds \leq 2\pi + 2 \qquad (\text{for } t=2^m, \; m=1,2,\ldots).$$

If $F(t) = \int_0^t f(s)ds$ is bounded for $t \geq 0$, then $F(t)$ is almost periodic. (This is Bohr's Theorem, but it also follows simply by applying Theorem 8 to the skew-product flow generated by $x' = f(t)$. In this case $\mathfrak{F}$ is the hull of f and the solution $\varphi(x,g,t)$ of $x' = g(t)$, $x(0) = x$ is simply

$$\varphi(x,g,t) = x + \int_0^t g(s)ds \quad .$$

The uniform stability is easily verified since

$$|\varphi(x_1,g,t) - \varphi(x_2,g,t)| = |x_1 - x_2| \quad .$$

If F is bounded for $t \geq 0$, then $\varphi(0,f,t) = F(t)$ is positively compact and therefore $\Omega(0,f)$ is a 1-cover of $\mathfrak{F}$ and is an almost periodic minimal set. Let $(x_1,f) \in \Omega(0,f)$. Since $\varphi(x_1,f,t)$ is almost periodic it follows that $F(t) = \varphi(x_1,f,t) - x_1$ is almost periodic. This implies, incidentally, that $x_1 = 0$, that is, $(0,f) \in \Omega(0,f)$.)

Next if $F(t)$ is almost periodic, then its Fourier series expansion would be the formal series

$$F(t) \sim \text{constant} + \sum_{k=1}^{\infty} \cos(2^{-k}\pi t) \ .$$

However the coefficients of $\cos(2^{-k}\pi t)$ do not have a finite ℓ_2 - sum and this violates Parseval's equality, [16]. Hence $F(t)$ is not bounded. Consequently the solution

$$\varphi(x, -f,t) = e^{-F(t)} x \qquad\qquad (x \neq 0)$$

of (3.5) is bounded for $t \geq 0$ but not uniformly stable. Furthermore no solution of (3.5), other than the trivial solution $\varphi(t) \equiv 0$, is almost periodic. Finally it follows from (3.6) that

$$|\varphi(x,-f,t)| \geq e^{-2\pi-2}|x| \ , \qquad\qquad \text{for} \quad t = 2^m, \ m = 1,2,\ldots \quad .$$

Then $\Omega(x,-f)$ is not an almost periodic minimal set. In fact, $\Omega(x,-f)$ is not even a minimal set in this example.

Remark 9. In the case of scalar equations, it follows from Appendix B that if M is a positively distal compact minimal set in $R^1 \times \mathfrak{F}$, then M is a 1-cover of $\mathfrak{F}$. More generally if M is a positively distal compact invariant set in $R^1 \times \mathfrak{F}$, then the flow π is distal on M and laminates into the union of 1-covers of $\mathfrak{F}$.

3.7 Triangularization of Linear Equations. One of the techniques that is oftentimes used in the study of linear differential equations $x' = A(t)x$ with time-varying coefficients is the method of triangularization. What one tries to do here is to find a change of variables $y = P(t)x$, where $P(t)$ is nonsingular and differentiable, such that the new equation in terms of the y-variable has an upper (or lower) triangular coefficient matrix. In other words, if $y' = B(t)y$ then

$$(3.7) \qquad B(t) = [P'(t) + P(t)\, A(t)]P^{-1}(t)$$

is to be upper (or lower) triangular. One of the first results on this technique is due to Perron [34], where it is shown that if $A(t)$ is bounded and continuous, then there is a bounded nonsingular matrix $P(t)$ with bounded inverse and such that $B(t)$ given by (3.7) is upper triangular. Also cf. Diliberto [8]. A more recent result due to Millionščikov [30] is that if the hull $H(A)$ is compact and minimal, then $P(t)$ can be chosen so that its hull $H(P)$ is compact and minimal. There is now a corresponding almost periodic problem. That is, if $A(t)$ is almost periodic, can one find an almost periodic $P(t)$ so that $B(t)$ given by (3.7) is upper triangular and almost periodic? One can show, by example, that it is not always possible to give a positive solution to this almost periodic problem, cf. Sacker and Sell [38]. However by using Theorem 2 one can give a sufficient condition for the existence of an almost periodic triangularization in the two-dimensional case.

We consider then a 2×2 system

$$(3.8) \qquad x' = A(t)x$$

where $x \in R^2$ and $A(t)$ is almost periodic in t. It is convenient to view $A(t)$ in a slightly different fashion. Let Y denote the hull of A and let $\sigma(y,t) = y \cdot t$ denote the corresponding flow on Y. That is, if $y = A^* \in Y$, then $y \cdot t = A_t^*$. It is known [45] that there is a continuous mapping Ψ from Y to the family of (2×2)-matrices such that for a fixed $y_o \in Y$ one has $\Psi(y_o \cdot t) = A(t)$. Equivalently, if one defines $A(y_o \cdot t)$ to be $A(t)$, then there is a unique continuous extension $A(y)$ defined for all $y \in Y$. We shall use $A(y)$ in place of $\Psi(y)$. Then $A(y)$ has the form

$$A(y) = \begin{bmatrix} a_{11}(y) & a_{12}(y) \\ a_{21}(y) & a_{22}(y) \end{bmatrix}$$

where each $a_{ij}(y)$ is a continuous real-valued function of $y \in Y$. Then equation (3.8) generalizes to

$$x' = A(y_o \cdot t)x \quad , \tag{3.9}$$

where y_o denotes some point in Y.

If we introduce polar coordinates (r,θ) in the plane R^2, then (3.9) can be written in the form

$$\theta' = a_{21}\cos^2\theta + (a_{22} - a_{11}) \sin\theta\cos\theta - a_{12} \sin^2\theta \tag{3.10}$$

$$r' = [a_{11}\cos^2\theta + (a_{12} + a_{21}) \sin\theta\cos\theta + a_{22} \sin^2\theta]\, r \tag{3.11}$$

The solutions of (3.10) introduce a skew-product flow $\pi = (\varphi,\sigma)$ on $S^1 \times Y$, where S^1 is the circle, $\sigma(y,t) = y \cdot t$, the coefficients

$a_{ij} = a_{ij}(y \cdot t)$ are evaluated along the curve $y \cdot t$ in Y and $\varphi(\theta,y,t)$ is the solution of (3.10) with $\varphi(\theta,y,0) = \theta$. A geometric condition for the existence of an almost periodic triangularization of (3.8) is now stated in terms of the skew-product flow on $S^1 \times Y$. Specifically, we ask for an N-fold covering of the base space Y.

Theorem 9. Let $A(t)$ be almost periodic in t and let Y denote the hull of A. Assume that the skew-product flow on $S^1 \times Y$ generated by (3.10) has a compact invariant set M that is an N-fold covering of the base space Y. Then there is an almost periodic nonsingular matrix $P(t)$ such that $P^{-1}(t)$, $P'(t)$ and $B(t)$, where $B(t)$ is given by (3.7), are almost periodic and, moreover, $B(t)$ is upper triangular.

Proof: Without any loss of generality, we can assume that M is a minimal set and therefore an almost periodic minimal set, cf. Theorem 2. Then the solution $\varphi(\theta,y,t)$ of (3.10), with $(\theta,y) \in M$, is almost periodic in t. Consequently the unit vector

$$e(t) = \begin{pmatrix} \cos(\varphi(\theta,y,t)) \\ \sin(\varphi(\theta,y,t)) \end{pmatrix}$$

is almost periodic in t. Let

$$f(t) = \begin{pmatrix} -\sin(\varphi(\theta,y,t)) \\ \cos(\varphi(\theta,y,t)) \end{pmatrix} .$$

Then $\{e(t), f(t)\}$ is an orthonormal set in R^2 for each $t \in R$, and these vectors are almost periodic functions of t. Let $P^{-1}(t) = \{e(t),f(t)\}$ be the (2×2)-orthogonal matrix formed by the column vectors $e(t)$ and

$f(t)$ respectively. Then clearly $P(t)$, $P'(t)$ and $P^{-1}(t)$ are almost periodic in t. Consequently the matrix $B(t)$ given by (3.7) is also almost periodic in t.

Now if $r(t)$ denotes the solution of (3.11), which is formed by replacing θ by $\varphi(\theta,y,t)$, and $r(0) = 1$, then $r(t)e(t)$ is a solution of (3.7). Consequently

$$y(t) = \begin{pmatrix} r(t) \\ 0 \end{pmatrix}$$

is a solution of $y' = B(t)y$. Since $r(t)$ is never zero this implies that $B(t)$ is upper triangular. (Also, cf. Diliberto [8].) Q.E.D.

Remark 10. There is an interesting relationship between the coefficient of the r-equation (3.11) and the coefficient of the linearized θ-equation. Specifically if we let $\theta_o(t)$ be a given solution of (3.10) and set $\theta = \theta_o(t) + \eta$, the linearized θ-equation is $\eta' = E(t)\eta$ where

$$E = -2(a_{12} + a_{21})\sin\theta_o \cos\theta_o + (a_{22} - a_{11})(\cos^2\theta_o - \sin^2\theta_o)$$

The r-equation (3.11) becomes $r' = R(t)r$ where

$$R = (a_{12} + a_{21})\sin\theta_o \cos\theta_o + a_{11}\cos^2\theta_o + a_{22}\sin^2\theta_o .$$

It is then easy to verify that

$$2R + E = \text{trace } A .$$

Chapter IV

Other Applications.

4.1 Retarded Differential Equations. Our general theory is applicable in the study of retarded differential equations with almost periodic coefficients. Let us briefly recall the essential features of this theory in order to imbed it in the abstract theory of skew-product semi-flows.

First we fix $r > 0$ and let $X = C([-r,0], R^n)$ denote the Banach space of continuous functions φ from $[-r,0]$ to R^n with the sup-norm. A continuous mapping $f: X \times R \to R^n$ then defines a retarded differential equation

$$x'(t) = f(x_t, t) \quad , \tag{4.1}$$

cf. Hale [19]. A solution of (4.1) on the interval $0 \leq t < a$ is defined to be a continuous function $\xi: [-r,a] \to R^n$ such that ξ is a C^1-function on $0 < t < a$, $\xi'(t) = f(\xi_t, t)$ for $t > 0$, where ξ_t is the profile of ξ at t and is defined by $\xi_t(\theta) = \xi(t+\theta)$ for $-r \leq \theta \leq 0$. Thus $\xi_t \in X$ for $0 \leq t < a$, and the "value" of ξ_t at $t = 0$ is the initial condition for the solution ξ, i.e., ξ_0 is an element of X.

The problems of the existence, uniqueness and continuity of solutions of retarded differential equations are similar to those for ordinary differential equations. However, one major difference in the two theories is that the solutions of retarded equations "prefer" positive time, i.e., the solutions of a retarded differential equation generate a semi-flow. In order to be more precise we let $\mathfrak{F}$ be a collection of continuous functions $f: X \times R \to R^n$ satisfying the following properties:

(H1) $\mathfrak{F}$ is translation - invariant, i.e., $f_\tau \in \mathfrak{F}$ whenever $f \in \mathfrak{F}$ and $\tau \in R$, where $f_\tau(\varphi,t) = f(\varphi, \tau + t)$.

(H2) Each function $f(\varphi,t)$ in $\mathfrak{F}$ is Lipschitz continuous in φ. (This implies that the initial value problem

$$x'(t) = f(x_t,t) \ , \ x_0 = \varphi$$

admits a unique solution ξ_t, which depends continuous on φ. We shall denote this solution by $T(\varphi,f,t)$ and we shall assume that $T(\varphi,f,t)$ is noncontinuable as a function of t.)

(H3) $\mathfrak{F}$ has a topology so that the two mappings

$$\sigma: \quad (f,\tau) \to f_\tau = \sigma(f,\tau)$$

$$T: \quad (\varphi,f,\tau) \to T(\varphi,f,\tau)$$

are continuous functions on their respective domains. For example, we may assume that $\mathfrak{F}$ has the topology of uniform convergence on bounded sets, cf. [19, p. 21].

Under these conditions on $\mathfrak{F}$ one can show that the mapping

$$\pi(\varphi,f,\tau) = (T(\varphi,f,\tau) \ , \ \sigma(f,\tau))$$

defines a skew-product semi-flow on $X \times \mathfrak{F}$. But notice that because of (H1), the mapping σ on the base space $\mathfrak{F}$ is actually a two-sided flow.

The question of the existence of compact positively invariant sets for the semi-flow π is somewhat more complicated than the corresponding problem for ordinary differential equations. In order to get a theory comparable to that described in Chapter III we shall make two further assumptions about $\mathfrak{F}$, viz.

(H4) $\mathfrak{F}$ is compact in the above topology, and

(H5) each $f \in \mathfrak{F}$ maps closed bounded sets in $X \times R$ into bounded sets in R^n .

If $\mathfrak{F}$ satisfies hypotheses (H1) - (H5) , then one can show that a semi-trajectory $\pi(\varphi,f,t)$ remains in a compact set in X for all $t \geq r$, if and only if there is a constant b such that $\|T(\varphi,f,t)\| \leq b$ for all $t \geq 0$ cf. [19, p.17] . This in turn implies that the ω-limit set $\Omega(\varphi,f)$ is a nonempty compact positively invariant set in the semi-flow π . Theorem 3 now stands ready for application.

As we argued in Section 3.3, the Amerio-Fink theory of separated solutions can now be applied (via Theorem 3) to retarded differential equations with almost periodic coefficients. (In this case $\mathfrak{F}$ is an almost periodic minimal set.) More precisely, if K is a compact set in X and there is a solution $T(\varphi,f,t)$ that remains in K for all $t \geq 0$ and the solutions of $x'(t) = f(x_t,t)$ are positively semi-separated in K , then $M = H^+(\varphi,f)$ is a compact positively invariant subset of the semi-flow π with the positive fibre-distal property. Furthermore the finiteness condition

$$\mu(f) = \text{card } p^{-1}(f) \cap M < \infty$$

is satisfied. Hence M is an N-fold covering space of the base $\mathfrak{F}$, the semi-flow π extends to a two-sided flow on M and M is an almost periodic minimal set.

The Miller-Yoshizawa theory of stable solutions described in Section 3.4 also can be extended to retarded differential equations but now one must exercise some care. First of all, Theorem 4, which says that uniform stabilities are inherited by the ω-limit set, is still valid. The proof of Theorem 4 given in [42] extends in a straight-forward manner to this more general situation. Next, one has the following extension of Theorem 5.

Theorem 10. *Let $\mathcal{F}$ be an almost periodic minimal set that satisfies properties* (H1) - (H5). *Assume that there is a positively compact solution* $T(\varphi,f,t)$ *that is uniformly stable. Then the* ω-*limit set* $\Omega(\varphi,f)$ *is a nonempty compact minimal set in* $X \times \mathcal{F}$. *Moreover the semi-flow* π *extends to a two-sided distal flow on* $\Omega(\varphi,f)$.

Proof: The main difficulty one encounters here is showing that the semi-flow on $\Omega(\varphi,f)$ can be extended to a two-sided flow. The remainder of the argument is then similar to the argument in Theorem 5. As a first step we restrict time t to take on integer values only. Our approach will be to construct a two-sided distal flow $\tilde{\pi}$ on a space $\tilde{M}$ which is an extension of the semi-flow π on $\Omega(\varphi,f)$. In other words, the flow $\tilde{\pi}$ on $\tilde{M}$ "projects" to the semi-flow π on $\Omega(\varphi,f)$. We shall then use the distal properties of $\tilde{\pi}$ to show that the semi-flow π can be extended to a two-sided flow.

Let $\tilde{X}$ denote the collection of all sequences $\xi(n)$, defined for $n \in Z$, with values in X. If d is the metric on X we construct a metric $\tilde{d}$ on $\tilde{X}$ by the formula

$$\tilde{d}(\xi,\hat{\xi}) = \Sigma_n 2^{-|n|} \min(1, d(\xi(n),\hat{\xi}(n))) .$$

It follows that $\min(1, d(\xi(n), \hat{\xi}(n))) \leq 2^{|n|} \tilde{d}(\xi,\hat{\xi})$ for all $n \in Z$. For each $h \in \mathcal{F}$ we define $\mathcal{S}(h)$ to be the set of all sequences $\xi \in \tilde{X}$ with the following two properties:

(i) $(\xi(n),h_n) \in \Omega(\varphi,f)$, for all $n \in Z$, and

(ii) $\pi(\xi(m),h_m,n) = (\xi(m+n), h_{m+n})$,

for all $m \in Z$ and all $n \in Z^+$. For each $\xi \in \tilde{X}$ we define the translate ξ_k by $\xi_k(n) = \xi(k+n)$. It is easy to verify that if $\xi \in \mathcal{S}(h)$, then

$\xi_k \in \mathcal{S}(h_k)$ for all $k \in Z$.

Through each point $(\psi,h) \in \Omega(\varphi,f)$ there exists a unique positive semi-orbit $\pi(\psi,h,t)$, $t \in Z^+$. This solution can be extended to $t \in Z^-$, perhaps not uniquely. More precisely, if $t_n \in Z^+$ is such that $t_n \to +\infty$ and $\pi(\varphi,f,t_n) \to (\psi,h)$, then for any $\tau \in Z^-$, $\pi(\varphi,f,t_n + \tau)$ contains a <u>subsequence</u> which converges to some $(\psi^*, h^*) \in \Omega(\varphi,f)$, and $\pi(\psi^*, h^*,-\tau) = (\psi,h)$. Each convergent subsequence gives rise to an extension, and each distinct extension gives rise to a distinct sequence $\xi(n)$ in the above construction. In particular for each $h \in \mathfrak{F}$, $\mathcal{S}(h)$ as defined above is nonempty. We now continue with the proof.

Next define $\tilde{M} = \{(\xi,h) \in \tilde{X} \times \mathfrak{F}: \xi \in \mathcal{S}(h)\}$ and define $\tilde{\pi}: \tilde{M} \times Z \to \tilde{M}$ by $\tilde{\pi}(\xi,h,n) = (\xi_n,h_n)$. Since $\Omega(\varphi,f)$ is compact it follows that $\tilde{M}$ is compact. Furthermore $\tilde{\pi}$ is a discrete two-sided flow on $\tilde{M}$. Define $\lambda: \tilde{M} \to M$ by $\lambda(\xi,h) = (\xi(0),h)$. Then λ is continuous and commutes with the flow for $n \geq 0$, i.e.

$$\pi(\xi(0),h,n) = \lambda(\tilde{\pi}(\xi,h,n)), \qquad n \geq 0 .$$

We claim that $\tilde{\pi}$ is negatively distal on M. In order to prove this we proceed by contradiction. Let $\delta(\nu)$ be the modulus of uniform stability for the solutions starting in $\Omega(\varphi,f)$. If $\tilde{\pi}$ were not negatively distal, then one could find two points (ξ,g), $(\hat{\xi},g) \in \tilde{M}$ so that $\xi \neq \hat{\xi}$ and $\inf_{n \leq 0} \tilde{d}(\xi_n,\hat{\xi}) = 0$. (We use here the fact that the flow on the base space $\mathfrak{F}$ is almost periodic or uniformly distal to conclude that the two points above have the same g-coordinate.) By replacing (ξ,g) and $(\hat{\xi},g)$ with translates if necessary we can assume that $\xi(0) \neq \hat{\xi}(0)$. Define $\nu_0 > 0$ by $2\nu_0 = d(\xi(0), \hat{\xi}(0))$. Let $\delta_0 =$ ~~$\delta(\nu_0)$~~ and choose $n < 0$ so that $\tilde{d}(\xi_n,\hat{\xi}_n) < \delta_0$ and therefore

min$(\delta(\nu_0), \frac{1}{2})$

$$d(\xi(n), \hat{\xi}(n)) = d(\xi_n(0), \hat{\xi}_n(0)) \leq \tilde{d}(\xi_n, \hat{\xi}_n) < \delta_o .$$

By the uniform stability one then has the contradiction

$$2\nu_o = d(\xi(0), \hat{\xi}(0)) = d(T(\xi(n), g_n, -n), T(\hat{\xi}(n), g_n, -n)) < \nu_o .$$

Hence the flow $\tilde{\pi}$ is negatively distal on $\tilde{M}$.

Now if N is any compact minimal set in $\tilde{M}$ then $\tilde{\pi}$ is negatively distal on N, and by Appendix B, $\tilde{\pi}$ is distal on N. This means that if (ξ, g), $(\hat{\xi}, g) \in N$ and $\xi \neq \hat{\xi}$ then there is an $\alpha > 0$ such that $d(\xi(n), \hat{\xi}(n)) \geq \alpha$ for all $n \in Z$. However one has

$$\xi(n) = T(\xi(0), g, n), \quad \hat{\xi}(n) = T(\hat{\xi}(0), g, n)$$

for $n \geq 0$. Since λ maps N onto a minimal set M in $\Omega(\varphi, f)$, this means that the discrete semi-flow π, when restricted to M, is one-to-one. Therefore $(\pi|M)^{-1}$ exists and is continuous. This shows that the restriction of the discrete semi-flow π to M can be extended to a two-sided discrete flow on M. It then follows that for $t \in R$, $t \geq 0$, the mapping $(\varphi, g) \to \pi(\varphi, g, t)$ is one-to-one on M, and therefore π can be extended to a two-sided continuous flow on M.

The rest of the argument now follows the argument of Theorem 5: The uniform stability on $\Omega(\varphi, f)$ implies that π is negatively distal on M. Hence π is distal on M. Therefore the mapping $p_1 = p|M$ is an open mapping of M onto the base space $\mathfrak{F}$. This them implies that $M = \Omega(\varphi, f)$. QED.

With this result behind us, the corresponding extension of Theorem 6 is now straightforward.

Theorem 11. *Let $\mathcal{F}$ be an almost periodic minimal set that satisfies properties (H1)-(H5). Assume that there is a positively compact solution $T(\varphi,f,t)$ that is uniformly asymptotically stable. Then the ω-limit set $\Omega(\varphi,f)$ is a nonempty compact almost periodic minimal set and is an N-fold covering space of $\mathcal{F}$, where N is finite. Moreover the semi-flow π extends to a two-sided flow on $\Omega(\varphi,f)$.*

Remarks 11. The last two theorems, which we formulated for retarded differential equations, are actually valid for skew-product semi-flows on $X \times Y$, where Y is an almost periodic minimal set (or more generally, a distal minimal set) in the induced flow σ. Our proofs carry over almost verbatim to this abstract setting.

12. Let f be an autonomous retarded differential equation and let $T(\sigma,f,t)$ be a positively compact uniformly stable solution of $x'(t) = f(x_t)$. Then $\Omega(\varphi,f)$ is a compact invariant set in the two-sided flow π, and π is uniformly stable on $\Omega(\varphi,f)$. By Markov's Theorem ([32, p. 390]), $\Omega(\varphi,f)$ is then an almost periodic minimal set. If, in addition, $T(\varphi,f,t)$ is uniformly asymptotically stable, then by Theorem 11, $\Omega(\varphi,f)$ consists of a single point, which must be a constant function in X and a zero of f.

There are other applications of our theory. The main idea is to construct a skew-product flow or semi-flow. Rather than going into great detail we just mention a few areas of application with some references to the literature.

4.2 Kurzweil Equations. In 1957 Kurzeil [23] introduced the notion of a generalized differential equation

$$x'(t) = D\,f(x(t),t)$$

and gave a theory concerning the existence of solutions for these equations.

The basic fact underlying this theory is that continuous nondifferentiable functions can be written as the limit of C^1-functions. This, in turn means that the derivatives of these C^1-functions converge in some weak sense. Kurzweil's theory places this in an appropriate structure and the generalized equations arise as limits of ordinary differential equations. Artstein [4] has shown that the Kurzweil equations can be imbedded in a skew-product flow.

4.3 Volterra Equations. The Volterra integral equation

$$x(t) = f(t) + \int_0^t a(t,s)\, g(x(s),s)ds$$

generates a skew-product semi-flow, cf. Miller and Sell [28]. In this case the base space Y consists of pairs $\{a,g\}$ where $a \in \mathcal{A}$, $g \in \mathcal{G}$, i.e., $Y = \mathcal{A} \times \mathcal{G}$. Uniform nonmetrizable topologies arise naturally in this context.

The Volterra differential-integral equation

$$x'(t) = f(x(t),t) + \int_0^t a(t,s)g(x(s),s)ds$$

also generates a skew-product semi-flow. (A discussion of the special case

$$x'(t) = A\,x(t) + f(t) + \int_0^t b(t-s)x(s)ds$$

can be found in [6,27]. The derivation of the skew-product semi-flow in the general case follows the methods of [28].)

Chapter V

Proofs of Theorems 1,2 and 3.

In this Chapter we shall present the proofs of the three theorems stated in Chapter III. In the case that both spaces X and Y are metriable, Theorems 1 and 2 are a special case of Sacker and Sell [36, Theorems 1 and 2]. The argument for the more general case, where Y is compact but not metrizable, is patterned after the argument in [36] but still requires more than the wave of a hand.

Before presenting the proof we should emphasize the importance of our assumption that the space X be metrizable and not merely uniform. Example 4 in [12] shows that the implication (C) => (A) in Theorem 1 can fail without some assumptions on the space X.

It will be convenient to simplify the notation. A point $w \in X \times Y$ will be denoted by $w = (x,y)$ where $x \in X$ and $y \in Y$. Thus w_1 and $\hat{w}$ will be $w_1 = (x_1, y_1)$ and $\hat{w} = (\hat{x}, \hat{y})$. The flow π on $X \times Y$ will be denoted by $w \cdot t$. The natural projection form $X \times Y$ to X will be denoted by q, thus $qw = x$. Recall that the natural projection from $X \times Y$ to Y is denoted by p.

If ρ is a pseudo-metric on Y and $\alpha > 0$ we define

$$B_{\rho,\alpha}(y) = \{\hat{y} \in Y: \rho(y,\hat{y}) < \alpha\}$$

$$B_{d\cdot\rho,\alpha}(w) = \{w \in X \times Y: d\cdot\rho(w,\hat{w}) < \alpha\}$$

where d is the metric on X and

$$d\cdot\rho(w,\hat{w}) = \max\{d(x,\hat{x}), \rho(y,\hat{y})\} .$$

If ρ and μ are pseudo-metrics on Y we say $\rho \leq \mu$ if for all $\alpha > 0$

and $y \in Y$ one has

$$B_{\rho,\alpha}(y) \subseteq B_{\mu,\alpha}(y) \quad .$$

Proof of Theorem 1: Let $M \subseteq X \times Y$ be a compact π-invariant set. We shall use p to denote the restriction $p|M$. Thus $p^{-1}(y)$ denotes the set $\{(x,y) \in M\}$.

(D) => (C). This is obvious.

(C) => (A). For each $\alpha > 0$ define

$$J_\alpha = \{y \in Y\colon \text{ if } w_1, w_2 \in p^{-1}(y) \text{ with } w_1 \neq w_2 \text{ , then } d(x_1,x_2) \geq \alpha\} \quad .$$

Notice that if $w_1, w_2 \in p^{-1}(y)$ then $d \cdot \rho(w_1,w_2) = d(x_1,x_2)$ for all pseudo-metrics ρ on Y. We claim that for every $\alpha > 0$, the set J_α is closed in Y. Indeed, let $\{y_n\}$ be a net or generalized sequence in J_α with $y_n \to y$. If $w_{1n}, w_{2n} \in p^{-1}(y_n)$ with $w_{1n} \neq w_{2n}$ then $d(x_{1n}, x_{2n}) \geq \alpha$. By choosing a subnet we may assume that $w_{1n} \to w_1$, $w_{2n} \to w_2$. Clearly one has $w_1, w_2 \in p^{-1}(y)$, and by the continuity of d, one has $d(x_1,x_2) \geq \alpha$. If we assume some fixed labeling of the N-points in $p^{-1}(y_n) = \{w_{1n}, \ldots, w_{Nn}\}$, then the nets $\{w_{1n}\}, \ldots, \{w_{Nn}\}$ have a common subnet converging to N distinct points $\{w_1, \ldots, w_N\}$ in $p^{-1}(y)$ with $d(x_i,x_j) \geq \alpha$ for $i \neq j$. Since these N points exhaust the fibre $p^{-1}(y)$ one has $y \in J_\alpha$. Hence J_α is closed.

Let $\{\alpha_n\}$ be a sequence of positive numbers with $\alpha_n \to 0$. Then $\bigcup_{n=1}^{\infty} J_{\alpha_n} = Y$. The next step in the argument is to show that $Y = J_{3\beta}$ for some $\beta > 0$. The proof of this fact, which is based on Baire's Theorem, is exactly the same as the argument in [36, p. 327]. Therefore we shall omit the details. Baire's Theorem is, of course, valid for the compact

Hausdorff space Y .

Remark 13. We have shown that if (C) holds then there is a $\beta > 0$ such that for all $y \in Y$ and for any pair $w_1, w_2 \in p^{-1}(y)$ with $w_1 \neq w_2$ one has

$$(5.1) \qquad d \cdot \rho(w_1 \cdot t, w_2 \cdot t) = d(q(w_1 \cdot t), q(w_2 \cdot t)) \geq 3\beta$$

for every pseudo-metric ρ and all $t \in T$. In other words, if (C) holds then $p = p|M$ is of distal type and the $\alpha(x_1, x_2, y)$ given in the definition of distal type can be chosen uniformly equal to 3β .

Remark 14. We have also shown that if (C) holds, then for any convergent net $\{y_n\}$ in Y with $y_n \to y$ one has $\limsup p^{-1}(y_n) = p^{-1}(y)$. In other words, $p = p|M$ is an open mapping. This is a special case of Theorem B.4 in Appendix B.

(C) => (D) . By using Remark 13 we fix $\beta > 0$ so that $d(x_1, x_2) \geq 3\beta$ whenever $(x_1, y), (x_2, y) \in M$ with $x_1 \neq x_2$. Now fix $y \in Y$. Then there is a pseudo-metric ρ on Y such that the N closed neighborhoods $\mathrm{Cl}\, B_{d \cdot \rho, \beta}(w_i)$ (where $1 \leq i \leq N$ and $p^{-1}(y) = \{w_1, \ldots, w_N\}$) are disjoint. By our choice of ρ , the restriction mapping $p_i = p|M \cap \mathrm{Cl}\, B_{d \cdot \rho, \beta}(w_i)$ is one-to-one for $1 \leq i \leq N$. Consequently each p_i is a homeomorphism of $M \cap B_{d \cdot \rho, \beta}(w_i)$ onto an open neighborhood of y . Since p is open, there exists (for $i = 1, \ldots, N$) an open subset $\mathcal{V}_i \subseteq Y$ such that

$$y \in \mathcal{V}_i \subseteq p_i(M \cap B_{d \cdot \rho, \beta}(w_i)) \quad .$$

Define $\mathcal{V} = \bigcap_{i=1}^{N} \mathcal{V}_i$ and $\mathcal{U}_i = p^{-1}(\mathcal{V}) \cap B_{d \cdot \rho, \beta}(w_i)$. Then $p^{-1}(\mathcal{V})$ is the disjoint union of $\mathcal{U}_1, \ldots, \mathcal{U}_N$, and for $1 \leq i \leq N$ and $p|\mathcal{U}_i = p_i|\mathcal{U}_i$ is a homeomorphism of $\mathcal{U}_i$ onto $\mathcal{V}$. Since this is valid for each $y \in Y$ we see that M is an N-fold covering space of Y with covering map $p = p|M$.

(A) => (B). This is obvious.

(B) => (C). Let y_o be given with $\mu(y_o) = N < \infty$ and let y be any other point in Y. Since Y is minimal the set $\gamma^+(y_o)$ is dense in Y, and therefore there is a net $\{t_n\}$ in T^+ with $y_o \cdot t_n \to y$. Let $p^{-1}(y_o) = \{w_1,\dots,w_N\}$. By the positive distal property there is an $\alpha > 0$ such that $d(q(w_i \cdot t), q(w_j \cdot t)) \geq \alpha$ for all $t \geq 0$, provided $i \neq j$. By choosing subnets, if necessary, we may assume that the sequences $\{w_1 \cdot t_n\},\dots,\{w_N \cdot t_n\}$ each converge to limits $\hat{w}_1,\dots,\hat{w}_N$. Clearly these limits are in $p^{-1}(y)$. Again by the continuity of d one has $d(q(\hat{w}_i), q(\hat{w}_j)) \geq \alpha$ for $i \neq j$. Therefore $\mu(y) = \text{card } p^{-1}(y) \geq N$. If $\mu(y) \geq N + 1$, then we use the fact that $\gamma^+(y)$ is dense in Y, and the same argument yields $\mu(y_o) \geq N + 1$, a contradiction. This completes the proof of the equivalence of the four statements (A), (B), (C) and (D).

The last assertion of Theorem 1, which describes M as the finite union of minimal sets, is proved in exactly the same fashion as in [36, pp 328-329]. Q.E.D.

<u>Proof of Theorem 2</u>; <u>Part (E)</u>: Let $w_1, w_2 \in M$ with $w_1 \neq w_2$. If $p(w_1) = p(w_2)$, then (5.1) implies that

$$d \cdot \rho(w_1 \cdot t, w_2 \cdot t) = d(q(w_1 \cdot t), q(w_2 \cdot t)) \geq 3\beta$$

for every pseudo-metric ρ and for all $t \in T$. On the other hand, if $p(w_1) \neq p(w_2)$ we define $y_i = p(w_i)$, $i = 1,2$. It follows from the distality of the flow σ on Y that there is a pseudo-metric ρ and an $\alpha > 0$ such that $\rho(y_1 \cdot t, y_2 \cdot t) \geq \alpha$ for all $t \in T$. Thus

$$d \cdot \rho(w_1 \cdot t, w_2 \cdot t) \geq \rho(y_1 \cdot t, y_2 \cdot t) \geq \alpha$$

for all $t \in T$. Hence π is distal on M.

Part (F): We will show that the flow π on M is uniformly distal. For this purpose we let $d\cdot\rho$ be a pseudo-metric on M and fix ν with $0 < \nu \leq \beta$, where β is determined by (5.1). Our objective is to find another pseudo-metric $d\cdot\hat{\rho}$ and a $\delta < 0$ so that

$$(5.2) \qquad [d\cdot\rho(w_1,w_2) \geq \nu, w_1,w_2 \in M] \Rightarrow [d\cdot\hat{\rho}(w_1\cdot t, w_2\cdot t) \geq \delta]$$

for all $t \in T$. The determination of $d\cdot\hat{\rho}$ and the verification of (5.2) will be done in a series of four lemmas.

Lemma 12. *There is a pseudo-metric ρ_1 on Y and an $\alpha_1 > 0$ with the following property: For every $y \in Y$, every pseudo-metric $\rho_2 \leq \rho_1$ and every positive $\alpha_2 \leq \alpha_1$ the set $p^{-1}(B_{\rho_2,\alpha_2}(y))$ consists of N disjoint open sets, which we label $W_{\rho_2,\alpha_2}(w_i)$, $1 \leq i \leq N$, where $p^{-1}(y) = \{w_1,\ldots,w_N\}$. Furthermore each set $W_{\rho_2,\alpha_2}(w_i)$ is homeomorphic to $B_{\rho_2,\alpha_2}(y)$ via the mapping p.*

Proof: We use the fact that M is an N-fold covering space of Y, and that Y is compact, to construct a finite open covering $\{V_1,\ldots,V_K\}$ of Y so that each set $p^{-1}(V_j)$ is the disjoint union of N open sets each of which is homeomorphic to V_j. One then chooses the pseudo-metric ρ_1 and the $\alpha_1 > 0$ so that for every $y \in Y$ the ball $B_{\rho_1,\alpha_1}(y)$ lies in at least one of the V_j, cf. Kelley [21, pp. 199-200] (In the metric case, the number α_1 is merely the Lebesgue number of the covering $\{V_1,\ldots,V_k\}$, Dugundji [9, p. 234].) If $\rho_2 \leq \rho_1$ and $\alpha_2 \leq \alpha_1$ then $B_{\rho_2,\alpha_2}(y) \subseteq B_{\rho_1,\alpha_1}(y)$ for every $y \in Y$ and the proof is complete.

We now fix the pseudo-metric ρ_1 and the number $\alpha_1 > 0$ by the above lemma. Let $\rho_2 \leq \rho_1$ and $0 < \alpha_2 \leq \alpha_1$. The sets

$W_{\rho_2,\alpha_2}(w_1),\dots,W_{\rho_2,\alpha_2}(w_N)$ will be called the W_{ρ_2,α_2}- levels of $B_{\rho_2,\alpha_2}(y)$ where $p^{-1}(y) = \{w_1,\dots,w_N\}$. Let $w,\hat{w} \in p^{-1}(B_{\rho_2,\alpha_2}(y^*))$ for some $y^* \in Y$. We shall say that w and $\hat{w}$ are in the same W_{ρ_2,α_2}- level if there is a $w_1 \in p^{-1}(y)$ such that $w,\hat{w} \in W_{\rho_2,\alpha_2}(w_1)$. Otherwise w and $\hat{w}$ are said to lie in different W_{ρ_2,α_2}- levels.

Lemma 13. *There is a pseudo-metric ρ_2 on Y and an $\alpha_2 > 0$ such that $\rho_2 \leq \rho_1$, $\alpha_2 \leq \alpha_1$ and for all pseudo-metrics $\rho_3 \leq \rho_2$ and all $\alpha_3 \leq \alpha_2$ $(\alpha_3 > 0)$ one has*

$$d(qw, q\hat{w}) > \beta$$

whenever w and $\hat{w}$ lie in different W_{ρ_3,α_3}- levels.

Proof: This follows from Lemma 12 and the uniform continuity of the mapping $w \to q\,w$.

We next fix the pseudo-metric ρ_2 and the number $\alpha_2 > 0$ by the above lemma. The notation $\rho_1 \vee \rho$, which we use below, is defined in Appendix A.

Lemma 14. *There is a pseudo-metric ρ_3 and a number $\alpha_3 > 0$ such that $\rho_3 \leq \rho_2 \vee \rho_1 \vee \rho$, $\alpha_3 < \min(\alpha_2, \nu)$ and with the property that if w and $\hat{w}$ lie in the same W_{ρ_3,α_3}- level then*

$$d(q(w \cdot t), q(\hat{w} \cdot t)) < \nu \leq \beta$$

for all $|t| \leq 1$. More generally, if for any $t_o \in T$, both $w \cdot t_o$ and $\hat{w} \cdot t_o$ lie in the same W_{ρ_3,α_3}- level then

$$d(q(w\cdot(t_o+t)),\ q(\hat{w}\cdot(t_o+t))) < \nu \le \beta$$

for all $|t| \le 1$.

Proof: This follows directly from the continuity of the flow π.

Now fix the pseudo-metric ρ_3 and the number $\alpha_3 > 0$ by the above lemma. The next lemma follows from the uniform distality of the flow σ on Y.

Lemma 15. There is a pseudo-metric $\rho_4 \le \rho_3$ and a number $\alpha_4 \le \alpha_3$ $(\alpha_4 > 0)$ such that for all $y, \hat{y} \in Y$ with $\rho_3(y,\hat{y}) \ge \alpha_3$ one has $\rho_4(y\cdot t,\ \hat{y}\cdot t) \ge \alpha_4$ for all $t \in T$.

We will now show that if $w, \hat{w} \in M$ then

$$(5.3) \qquad [d\cdot\rho(w,\hat{w}) \ge \nu] \Rightarrow [d\cdot\rho_4(w\cdot t,\ \hat{w}\cdot t) \ge \alpha_4]$$

for all $t \in T$, where ρ_4 and α_4 are given by the above lemma. In other words, the flow π is uniformly distal on M.

The verification of (5.3) will be done in each of three cases:

Case 1. $\rho_3(y\cdot t_o,\ \hat{y}\cdot t_o) \ge \alpha_3$ for some $t_o \in T$.

In this case it follows from Lemma 15 that

$$d\cdot\rho_4(w\cdot t,\ \hat{w}\cdot t) \ge \rho_4(y\cdot t,\ \hat{y}\cdot t) \ge \alpha_4$$

for all $t \in T$, i.e. (5.3) is valid.

The next two cases deal with points $w, \hat{w} \in M$ with the property that $\rho_3(y\cdot t,\ \hat{y}\cdot t) < \alpha_3$ for all $t \in T$, i.e. with $w\cdot t,\ \hat{w}\cdot t \in p^{-1}(B_{\rho_3,\alpha_3}(y\cdot t))$ for all $t \in T$. We then distinguish between whether or not $w\cdot t$ and $\hat{w}\cdot t$ lie on the same W_{ρ_3,α_3} - level.

Case 2. $\rho_3(y \cdot t, \hat{y} \cdot t) < \alpha_3$ for all $t \in T$ and $w \cdot t$ and $\hat{w} \cdot t$ lie in different W_{ρ_3,α_3}- levels for all $t \in T$.

Since $w \cdot t$ and $\hat{w} \cdot t$ lie in different W_{ρ_3,α_3}- levels for all $t \in T$ it follows from Lemma 13 that $d(q(w \cdot t), q(\hat{w} \cdot t)) > \beta$ for all $t \in T$. Hence

$$d\cdot\rho_4(w\cdot t,\hat{w}\cdot t) \geq d(q(w,t), q(\hat{w},t)) > \beta \geq \alpha_4$$

for all $t \in T$, i.e. (5.3) is valid.

Case 3. $\rho_3(y\cdot t, \hat{y}\cdot t) < \alpha_3$ for all $t \in T$ and $w\cdot t_o$ and $\hat{w}\cdot t_o$ lie in the same W_{ρ_3,α_3} - level for some $t_o \in T$.

The argument will be completed by showing that this case leads to a contradiction. Consider the inequality $\rho_3(y\cdot t, \hat{y}\cdot t) < \alpha_3 < \nu$ at $t = 0$. The inequality $\rho_3 \leq \rho$ then implies that $\rho(y,\hat{y}) < \nu$. On the other hand

$$d\cdot\rho(w,\hat{w}) = \max\{d(qw,q\hat{w}), \rho(y,\hat{y})\} \geq \nu$$

implies that $d(qw,q\hat{w}) \geq \nu$. Lemma 14 in turn implies that w and $\hat{w}$ lie in different W_{ρ_3,α_3}- levels. Hence $t_o \neq 0$. (We will now assume that $t_o > 0$. A similar argument will handle the case where $t_o < 0$.) Next choose $w_1 \in p^{-1}(y)$ so that $\hat{w}$ and w_1 lie on the same W_{ρ_3,α_3}- level. Then of necessity one has $w_1 \neq w$. Since $w,w_1 \in p^{-1}(y)$, (5.1) implies that $d\cdot\rho_4(w_1\cdot t, w\cdot t) \geq 3\beta$ for all $t \in T$. Since $w\cdot t_o$ and $\hat{w}\cdot t_o$ lie in the same W_{ρ_3,α_3}- level and since $w\cdot t_o \neq w_1\cdot t_o$, it follows that $w_1\cdot t_o$ and $\hat{w}\cdot t_o$ lie in different W_{ρ_3,α_3}- levels. Define

t_1 as the infimum of the set of $t>0$ such that $w_1 \cdot t$ and $\hat{w} \cdot t$ lie in different W_{ρ_3, α_3} - levels. Then choose t_2 and τ so that $0 \leq t_2 \leq t_1$, $0 < \tau \leq 1$ and for $t = t_2 + \tau$ one has

(5.4) $w_1 \cdot t_2$ and $\hat{w} \cdot t_2$ lie in the same W_{ρ_3, α_3} - level, and

(5.5) $w_1 \cdot t$ and $\hat{w} \cdot t$ lie in different W_{ρ_3, α_3} - levels.

Then Lemma 14 and (5.4) imply that

$$d(q(w_1 \cdot t),\ q(\hat{w} \cdot t)) < \nu \leq \beta ,$$

while Lemma 13 and (5.5) imply that

$$d(q(w_1 \cdot t),\ q(\hat{w} \cdot t)) > \beta ,$$

a contradiction. QED.

Proof of Theorem 3: The argument here is similar to the proof of Theorem 1. (D) => (C) is obvious. For implication (C) =>(A), one first shows that $Y = J_{3\beta}$ for some $\beta > 0$, by using the argument of Theorem 1. This then implies that the semi-flow $\pi_t \colon M \to M$ is one-to-one for $t \geq 0$. Since M is compact this means that $(\pi_t)^{-1}$ exists. For $t < 0$ define $\pi_t = (\pi_{-t})^{-1}$. This extends π to a continuous mapping from $M \times T$ to M. It is then easy to verify the group property for this extension. Consequently this extension defines a two-sided flow. The remaining implications (C)=> (D), (A) => (B) and (B) => (C) are now verbatim copies of Theorem 1. Q.E.D.

Chapter VI

An Example.

One may ask whether the assumption that

$$\mu(y) = \text{card } p^{-1}(y) \cap M$$

be finite for all $y \in Y$ implies that $\mu(y)$ is constant. The answer is no. In the case that M is an "extension of Y", as described in Sacker and Sell [36], an example due to Veech [47], shows that the finiteness of $\mu(y)$ does not imply constancy. One can show that the Veech example can be suitably imbedded in a skew-product flow. But rather than doing this, let us turn instead to a differential equation constructed by Kato and Sibuya [20], which is instructive in other matters.

The example starts with a continuous function $F(x,y_1,y_2): R^3 \to R^1$ that is doubly-periodic in y_1,y_2, i.e.

$$F(x,y_1,y_2) = F(x,y_1 + 1,y_2) = F(x,y_1,y_2+1)$$

for all x,y_1,y_2. In other words, F can be considered as a mapping of $R^1 \times T^2$ into R^1, where T^2 is the two-dimensional torus. Next, an irrational number α is chosen and one considers the collection of all differential equations

$$x' = F(x,y_1 + t, y_2 + \alpha t) = f(x,y \cdot t), \tag{6.1}$$

where $y = (y_1,y_2)$ denotes an arbitrary point in T^2. The function f is then quasi-periodic and the hull $\mathfrak{F}$ is precisely the torus T^2. The translational flow $\sigma(f,\tau) = f_\tau$ is then the irrational twist flow on T^2 given by $y \cdot \tau = (y_1 + \tau, y_2 + \alpha\tau)$. The chosen function F satisfies

$|F| \leq A|x| + B$, for constants A and B so that all solutions of (6.1) are defined for all $t \in R$.

The function F is constructed with the property that if (y_1,y_2) does not lie on the trajectory in T^2 passing through $(0,0)$ then the solutions of (6.1) are unique, bounded for $t \geq 0$ and uniformly asymptotically stable. The differential equation (6.1) that corresponds to (y_1,y_2) on the trajectory through $(0,0)$ does not have unique solutions.

In order to construct a flow for this example one uses the methods described in Section 3.5 and [44]. That is, for each $y \in T^2$ let $\mathcal{S}(y)$ denote the collection of all noncontinuable solutions of $x' = f(x, y \cdot t)$. Then $\mathcal{S}(y) \subseteq C$ for all $y \in T^2$, where $C = C(R,R)$ is the space of continuous functions from R to R . Let

$$W = \{(\varphi,y) \in C \times T^2 : \varphi \in \mathcal{S}(y)\}$$

then $\pi(\varphi,y,\tau) = (\varphi_\tau, y\cdot\tau)$ is a global flow on W . As noted in Section 3.5, W is an invariant set for a skew-product flow on $C \times T^2$. The example, in addition, has the following properties:

1. There is precisely one ω-limit set M in this flow on W . Furthermore M is nonempty, compact and minimal.

2. For $y = (y_1,y_2)$ not on a trajectory through $(0,0)$ in T^2 one has

$$\mu(y) = \text{card}\,\{(\varphi,y) \in M\} = 1$$

3. For $y = (y_1,y_2)$ on the trajectory through $(0,0)$ in T^2 one has $\mu(y) = 2$.

This example also shows that in the case of nonuniqueness, the uniform stability properties are not inherited by the limiting equations. The reason for this is connected with the fact that the fibres $\mathcal{S}(y)$ are not homeomorphic. Indeed, if $y = (y_1, y_2)$ is not a trajectory through $(0,0)$ then the fibre $\mathcal{S}(y)$ is homeomorphic to R^1. The other fibres are far more complicated, [44].

Chapter VII

Flows on Fibre Bundles.

The basic theory described in Theorems 1 and 2, as well as the stability Theorems 4, 5 and 6, can be extended to fibre-preserving flows on fibre bundles.

A topological Hausdorff space W is said to be a locally trivial fibre bundle or simply a fibre bundle with base space Y, fibre X and projection p if the following hold:

(i) W, Y and X are topological Hausdorff spaces,

(ii) $p: W \to Y$ is a continuous mapping of W onto Y, and

(iii) for each $y \in Y$ there is a neighborhood U of y such that $p^{-1}(U)$ is homeomorphic to the product space $X \times U$, i.e., there exists a homeomorphism $\Phi: X \times U \to p^{-1}(U)$ such that for each $y \in U$, the restriction of Φ to $X \times \{y\}$ is a homeomorphism of $X \times \{y\}$ onto $p^{-1}(y)$.

In the usual literature a fibre bundle has an additional structure, which is not required in our theory, so we delete it. The additional structure is described in terms of a fixed group G of homeomorphisms of the fibre X onto, cf. [46] for details.

In order for our theory to apply we shall require the fibre X to be metrizable and the base space Y to be a compact space. However for simplicity we shall assume that all three spaces W, Y and X are metrizable. We shall let d denote the metric on W.

Let W be a fibre bundle with compact base space Y, fibre X, and a projection p. We shall say that a flow π on W is a skew-product flow if there is a flow σ on Y such that

$$p(\pi(w,t)) = \sigma(p(w),t)$$

for all $w \in W$ and all t. In the language of flows, this means that p is a homomorphism of transformation groups, i.e., p commutes with the flow.

Property (iii) in the definition of a fibre bundle allows one to introduce local coordinates on W. A skew-product flow can then be described in terms of these local coordinates. More precisely, let U be an open set in Y with the property that both $p^{-1}(U)$ and $p^{-1}(U\cdot t)$ are homeomorphic to $X \times U$ and $X \times U\cdot t$, respectively. (Here we write $U\cdot t$ for $\sigma(U,t)$.) Let

$$\Phi: p^{-1}(U) \to X \times U$$

$$\Phi^t: p^{-1}(U\cdot t) \to X \times U\cdot t$$

be the homeomorphisms described by (iii). Then for each $(x,y) \in X \times U$ one has $w = \Phi^{-1}(x,y) \in p^{-1}(U) \subseteq W$ and $\pi(w,t) \in p^{-1}(U\cdot t)$. Then $\Phi^t(\pi(w,t)) \in X \times U\cdot t$. Since p is a homomorphism one has

$$\Phi^t(\pi(w,t)) = (\varphi(x,y,t) \ , \ y\cdot t)$$

where $\varphi(x,y,t) = \Phi^t \circ \pi_t \circ \Phi^{-1}(x,y)$, $y\cdot t = \sigma(y,t)$ and $\pi_t(w) = \pi(w,t)$. The function $\varphi(\cdot,y,t)$ is then a mapping from X to X which is defined for all $y \in U$. Clearly it depends on the homeomorphisms Φ and Φ^t.

In simple terms then, a fibre bundle W is a topological space that looks locally like a product space $X \times U$, where U varies over a collection of small neighborhoods in Y. A skew product flow on W is then a flow that has a local representation of the form

$$\pi(x,y,t) = (\varphi(x,y,t) \ , \ y\cdot t)$$

where $y\cdot t = \sigma(y,t)$ is a flow on Y.

Let π be a skew-product flow on a fibre bundle W with compact base space Y, fibre X and projection p. Let M be a compact π-invariant set in W. One says that $p|M$, the restriction of p to M, is of distal type (and that the invariant set M has the fibre-distal property) if for any two points w_1, w_2 in $p^{-1}(y) \cap M$ with $w_1 \neq w_2$ there is an $\alpha = \alpha(w_1, w_2) > 0$ such that $d(\pi(w_1,t), \pi(w_2,t)) \geq \alpha$ for all $t \in T$. Because of the assumption of the metrizability of W, Y and X, Theorems 1 and 2 follow from [36, Theorems 1 and 2]. What is of greater interest in this context are the Stability Theorems 4,5 and 6. The proper generalization of the stability concepts is a conditional stability which we refer to as fibre-wise stability.

A motion $\pi(w,t)$ is said to be fibre-wise uniformly stable if $\pi(w,t)$ is defined for all $t \geq 0$ and for every $\nu > 0$ there is a $\delta = \delta(\nu) > 0$ such that

$$d(\pi(w,\tau+t), \pi(\hat{w},\tau+t)) \leq \nu, \qquad \text{(for all } t \geq 0)$$

whenever $\hat{w} \in p^{-1}(p(w))$, $\tau \geq 0$ and $d(\pi(w,\tau), \pi(\hat{w},\tau)) \leq \delta$. The motion $\pi(w,t)$ is said to be fibre-wise uniformly asymptotically stable if it is fibre-wise uniformly stable and there is a $\delta_0 > 0$ and for every $\nu > 0$ there is a $t_0 > 0$ such that

$$d(\pi(w,\tau+t), \pi(\hat{w},\tau+t)) \leq \nu, \qquad \text{(for all } t \geq t_0)$$

whenever $\hat{w} \in p^{-1}(p(w))$, $\tau \geq 0$ and $d(\pi(w,\tau), \pi(\hat{w},\tau)) \leq \delta_0$.

Theorem 4 now extends to skew-product flows on fibre bundles with a compact base space provided one replaces the stability concepts with the fibre-wise stability concepts defined above. The proof of Theorem 4 given in [43] extends directly to this more general context when one uses the local coordinate representation of motion in W described above. Theorems 5 and 6 then have the following extensions:

Theorem 16. Let π *be a skew-product flow on a fibre bundle* W *with compact base space* Y. *Assume that* Y *is an almost periodic minimal set in the induced flow* σ.

(A) *If there is a positively compact motion* $\pi(w,t)$ *that is fibre-wise uniformly stable, then the* ω*-limit set* $\Omega(w)$ *is a nonempty compact minimal set in* W *and the flow* π *is distal on* $\Omega(w)$.

(B) *If there is a positively compact motion* $\pi(w,t)$ *that is fibre-wise uniformly asymptotically stable, then the* ω*-limit set* $\Omega(w)$ *is an N-fold covering space of* Y, *where* N *is finite, and* $\Omega(w)$ *is an almost periodic minimal set.*

Skew-product flows on a fibre bundle W with fibre $X = R^1$ are particularly interesting. This is the fibre-bundle generalization of the theory of scalar equations. Theorem 8 now takes the following form:

Theorem 17. Let π *be a skew-product flow on a fibre bundle* W *with compact base space* Y *and fibre* $X = R^1$. *Assume that* Y *is an almost periodic minimal set in the induced flow* σ. *If there is a positively compact motion* $\pi(w,t)$ *that is uniformly stable, then the* ω*-limit set* $\Omega(w)$ *is an N-fold covering of the base space* Y, *where* $N = 1$ *or* 2. *In particular* $\Omega(w)$ *is an almost periodic minimal set.*

Proof: It follows from Theorem 16 that the flow π is distal on $M = \Omega(w)$. Because of Theorems 1 and 2 it will suffice to show that no fibre $p^{-1}(y) \cap M$ contains more than two-points, and we do this by contradiction. Assume there is a $y \in Y$ such that $p^{-1}(y) \cap M$ has at least three points, call them w_1, w_2 and w_3. Since $p^{-1}(y) \cap M$ is compact and the fibre $p^{-1}(y)$ is homeomorphic to R^1 we can place a total ordering on $p^{-1}(y)$ and chose the three points so that

$$w_1 = \inf \{w \in p^{-1}(y) : w \in M\}$$

$$w_3 = \sup \{w \in p^{-1}(y) : w \in M\}$$

and $w_1 < w_2 < w_3$. Now use the minimality and the compactness of M to choose a sequence $\{\tau_n\}$ in T so that $\pi(w_2, \tau_n) \to w_1$ and $\pi(w_1, \tau_n)$ and $\pi(w_3, \tau_n)$ converge. Because of the definition of w_1 and w_3 one has either (i) $\pi(w_1, \tau_n) \to w_1$ or (ii) $\pi(w_3, \tau_n) \to w_1$. Either case contradicts the distality of the flow on M and consequently M is an N-fold covering of Y with $N = 1$ or 2 . The remainder of the proof follows from Theorem 2. QED.

Appendix A

Uniform Spaces

Let X be a nonempty set and let $\mathcal{P} = \{\rho\}$ be a family of pseudo-matrics on X. The family $\mathcal{P}$ generates a topology on X as follows: A set $A \subseteq X$ is said to be open if for every $x \in A$ there is a finite collection $\rho_i \in \mathcal{P}$ and $\alpha_i > 0$, $1 \leq i \leq n$ such that

$$\cap_{i=1}^{n} B_{\rho_i,\alpha_i}(x) \subseteq A$$

where $B_{\rho,\alpha}(x) = \{\hat{x} \in X: \rho(x,\hat{x}) < \alpha\}$. The collection of all open sets is then the topology $\mathcal{T}(\mathcal{P})$ on X. One can show that $(X,\mathcal{T}(\mathcal{P}))$ is a Hausdorff space if and only if given any two distinct points $x,y \in X$ there is a $\rho \in \mathcal{P}$ such that $\rho(x,y) > 0$. A topological space $(X,\mathcal{T}(\mathcal{P}))$, with the topology generated in the above fashion, is called a uniform space. (Uniform spaces arise in other equivalent manners. For a complete discussion, cf. Dugundji [9] or Kelley [21].)

If ρ and σ are two pseudo-metrics on X, then $\rho \vee \sigma = \max(\rho,\sigma)$ is also a pseudo-metric on X. Let $\mathcal{P} = \{\rho\}$ be a family of pseudo-metrics on X and let $\mathcal{P}'$ denote the smallest collection of pseudo-metrics on X satisfying (i) $\mathcal{P} \subseteq \mathcal{P}'$ and (ii) $\rho \vee \sigma \in \mathcal{P}'$ whenever $\rho, \sigma \in \mathcal{P}'$. One can easily show that $\mathcal{T}(\mathcal{P}) = \mathcal{T}(\mathcal{P}')$. For this reason we assume that the family of pseudo-metrics $\mathcal{P}$ is closed under the operation $\vee$.

Let ρ and σ be two pseudo-metrics on X. We shall write $\rho \leq \sigma$ if for every $\nu > 0$ one has $\sigma(x,y) < \nu$ whenever $\rho(x,y) < \nu$. This means that for every $x \in X$ one has $B_{\rho,\nu}(x) \subseteq B_{\sigma,\nu}(x)$. Notice that for any two pseudo-metrics ρ and σ one always has $\rho \vee \sigma \leq \rho$ and $\rho \vee \sigma \leq \sigma$. In particular, if $\mathcal{P}$ is a family of pseudo-metrics on X that is closed under $\vee$, then for any two pseudo-metrics ρ_1 and ρ_2 in $\mathcal{P}$, there is a third pseudo-metric ρ_3 such that $\rho_3 \leq \rho_1$ and $\rho_3 \leq \rho_2$.

For our purposes we shall need the following fact:

Theorem. Every compact Hausdorff space is a uniform space.

The proof is not difficult. One is given a compact Hausdorff space $(X,\mathcal{T})$ with a topology $\mathcal{T}$, and one needs to determine a family $\mathcal{P}$ of pseudo-metrics on X so that $\mathcal{T} = \mathcal{T}(\mathcal{P})$. In order to do this we let $C(X,R)$ denote the collection of all continuous real-valued functions on X . For each $f \in C(X,R)$ we define a pseudo-metric by

$$\rho^f(x,y) = |f(x) - f(y)| .$$

Let $\mathcal{P} = \{\rho^f: f \in C(X,R)\}$ and determine $\mathcal{T}(\mathcal{P})$ as described above. One can then show that $\mathcal{T} = \mathcal{T}(\mathcal{P})$. The inequality $\mathcal{T}(\mathcal{P}) \subseteq \mathcal{T}$ follows from the characterization of continuity in terms of the inverse images of open sets. The opposite inequality $\mathcal{T} \subseteq \mathcal{T}(\mathcal{P})$ follows from the fact that $(X,\mathcal{T})$ is completely regular, i.e., given a point $x_o \in X$ and a closed subset $A \subseteq X$, not containing x_o , then there is a continuous $f: X \to [0,1]$ such that $f(x_o) = 0$ and $f(x) = 1$ for all $x \in A$.

Appendix B

Distal Flows and the Enveloping Semigroup

The object of this appendix is to prove two theorems concerning distal flows. The first shows that one-sided distality is equivalent to two-sided distality.

Theorem B.1. Let π be a negatively distal flow on a compact Hausdorff space W. Then π is distal.

By the Tychonoff Theorem the set W^W of all mappings (continuous or not) of W into W is a compact space with the product topology of pointwise convergence. The set W^W is naturally provided with a semi-group structure under composition of functions, i.e., $fg = f \circ g$ where $f \circ g(w) = f(g(w))$.

Lemma B.2 (A). Right multiplication is continuous, i.e., for each $q \in W^W$, $R_q: p \to pq$ is continuous at each $p \in W^W$.

(B) Left multiplication $L_q: p \to qp$ is continuous at each $p \in W^W$ whenever q is itself a continuous mapping of W into W.

Proof: (A) Let p_i be a generalized sequence in W^W, with $p_i \to p$, and let $q \in W^W$. Then for each $w \in W$, $q(w) \in W$ and therefore $p_i(q(w)) \to p(q(w))$, i.e., $p_i q(w) \to p\,q(w)$. Hence R_q is continuous at p.

(B) If q is continuous then $q(p_i(w)) \to q(p(w))$. That is, $q\,p_i(w) \to q\,p(w)$. Hence L_q is continuous when q is continuous. QED.

For each $t \in T$ let $\pi_t(w) = \pi(w,t)$. Then $\pi_t \in W^W$. Define Γ to be the closure of $\{\pi_t: t \in T\}$ in W^W and Γ^- to be the closure of $\{\pi_t: t \in T, t \leq 0\}$. Then Γ and Γ^- are compact subsets of W^W. A typical element in Γ^- is of the form

(B.1) $$\gamma(w) = \lim \pi_{t_i}(w)$$

for some generalized sequence $t_i \leq 0$. Elements of Γ are described similarly with no ordering restriction on the generalized sequence t_i.

Lemma B.3 Γ and Γ^- are sub-semigroups of W^W.

Proof: We shall prove it for Γ. Let γ and $\mu \in \Gamma$ and let $\gamma_i \to \gamma$ and $\mu_j \to \mu$, where $\gamma_i = \pi_{t_i}$ and $\mu_j = \pi_{s_j}$ are generalized sequences. Now fix i. Since each γ_i is continuous it follows from Lemma B.2 that $\lim_j \gamma_i \mu_j = \gamma_i \mu$. Once again Lemma B.2 implies that $\lim_i \gamma_i \mu = \gamma \mu$, hence $\gamma \mu \in \Gamma$. QED.

What is the connection between distality and these semi-groups? The answer is simple and elegant.

Lemma B.4 The discrete flow π is negatively distal on W if and only if every $\gamma \in \Gamma^-$ is one-to-one. Similarly, π is distal on W if and only if every $\gamma \in \Gamma$ is one-to-one.

The proof of Lemma B.4 is rather straightforward. Assume that the flow π is negatively distal and let $\gamma \in \Gamma^-$ be chosen so that (B.1) holds. It is clear that if $\gamma(w_1) = \gamma(w_2)$ one then has

$$\lim \pi_{t_i}(w_1) = \lim \pi_{t_i}(w_2) , \tag{B.2}$$

which implies that $w_1 = w_2$ by the negative distality. Conversely assume that every $\gamma \in \Gamma^-$ is one-to-one. If $w_1, w_2 \in W$ are chosen so that (B.2) holds for some generalized sequence $t_i \leq 0$, then one can chose a generalized subsequence - call it again t_i - so that $\pi_{t_i} \to \gamma \in \Gamma^-$. Equation (B.2) then implies that $\gamma(w_1) = \gamma(w_2)$, which in turn implies that $w_1 = w_2$. Hence π is negatively distal. The argument concerning the distality of π is similar.

Next we claim that if the flow π is negatively distal on W then one has the left cancellation law

$$\text{(B.3)} \qquad [\gamma\gamma_1 = \gamma\gamma_2] \Rightarrow [\gamma_1 = \gamma_2]$$

for all $\gamma, \gamma_1, \gamma_2 \in \Gamma^-$. This follows simply from the fact that γ is one-to-one.

Lemma B.5 *Assume that the flow* π *is negatively distal on* W. *Then for every* $\gamma \in \Gamma^-$ *there is* $\psi \in \Gamma^-$ *so that* $\psi\gamma = e$, *where* $e : W \to W$ *is the identity mapping*.

If Lemma B.5 is valid, then by setting $\gamma = \pi_t$ for some $t < 0$, we see that $\psi = \pi_{-t} \in \Gamma^-$. This means that $\{\pi_t : t \in T\} \subset \Gamma^-$, which implies that $\Gamma \subseteq \Gamma^-$. By Lemma B.4 we conclude that every $\gamma \in \Gamma$ is one-to-one and hence the flow π on W is distal. Thus the proof of Theorem B.1 reduces to proving Lemma B.5.

In order to prove Lemma B.5, we fix a $\gamma \in \Gamma^-$. Now let $E = \Gamma^- \gamma = \{\psi\gamma : \psi \in \Gamma^-\}$. Since R_γ is continuous and Γ^- is compact, we see that E is compact. Furthermore E itself is a semigroup. Next we claim that E has an idempotent, i.e. there is a $u \in E$ such that $u^2 = u$. If u is an idempotent in E, then there is a $\psi \in \Gamma^-$ so that $u = \psi\gamma$. Now $u^2 = u$ becomes

$$\psi\gamma\psi\gamma = \psi\gamma = \psi\gamma e \quad .$$

But then (B.3) implies that $u = \psi\gamma = e$. Thus it remains to show that E has an idempotent.

Let $\mathcal{A}$ denote the collection of all nonempty compact sets $A \subseteq E$ such that $A^2 \subseteq A$. This collection is nonempty since $E \in \mathcal{A}$. Let $\mathcal{A}$

be ordered by inclusion. It is easy to verify that Zorn's Lemma applies, and therefore $\mathcal{Q}$ contains a minimal element B. Let $b \in B$. Then Bb is a nonempty compact set in E since R_b is continuous. Furthermore $Bb \subseteq BB \subseteq B$. Also one has $(Bb)^2 \subseteq B^2 b \subseteq Bb$. Hence $Bb \in \mathcal{Q}$ and by minimality $Bb = B$. Therefore there is an $a \in B$ such that $ab = b$. Now define A by

$$A = \{a \in B: ab = b\} .$$

Since $A = R_b^{-1}(b) \cap B$ we see that A is compact. It is also nonempty. Furthermore one has $A^2 \subseteq A$. Since $A \subseteq B$, the minimality of B means $A = B$. Thus $b \in A$, or $b^2 = b$. Q.E.D.

Remark 15. The above argument is based on the theories of Ellis [11] and Furstenberg [17]. The set Γ we construct above is precisely the "enveloping semigroup" used in these theories. Note that it was the one-to-oneness property of elements of Γ^- which made the proof of Lemma B.5 work. Once it is known that $\Gamma \subseteq \Gamma^-$ then a similar lemma applies to Γ, i.e., every $\gamma \in \Gamma$ has a left inverse $\psi \in \Gamma$. But ψ also has a left inverse $\alpha \in \Gamma$, and hence $\alpha \psi = \psi \gamma = e$; from which it follows that $\alpha = \gamma^{-1}$. In other words, Γ *is a group whenever the flow* π *on* W *is distal*. However, Γ need not be a topological group since the group operations will not in general be continuous.

There is a deeper connection between almost periodicity and Γ. In fact, the flow π is almost periodic on W if and only if Γ is a group of continuous mappings of W onto W, cf. Ellis [11, p. 25-26].

The second theorem we consider is

Theorem B.4. *Let* π *and* σ *be flows on compact Hausdorff spaces* W *and* Y *respectively, and assume that* W *and* Y *are minimal in these flows.*

Assume further that

(i) the flow σ is distal on Y, and

(ii) there is a homomorphism $p: W \to Y$, i.e. a continuous mapping of W onto Y that commutes with the flows, and

(iii) p is of distal-type, i.e. given $w_1, w_2 \in p^{-1}(y)$ for any $y \in Y$ and $w_1 \neq w_2$, then there is a pseudo-metric ρ on W and an $\alpha > 0$ so that $\rho(\pi(w_1,t), \pi(w_2,t)) > \alpha$ for all $t \in T$.

Then the flow π on W is distal and the mapping p is open.

The distality of the flow π is easily verified. (Two points in W lying in the same fibre stay apart by the fibre distal assumption while two points lying in different fibres stay apart due to the distality of Y.) Now let Γ denote the closure of $\{\pi_t: t \in T\}$ in W^W. Then as seen above Γ is a group. In particular each element of Γ is a one-to-one mapping of W onto W. Hence for each $\gamma \in \Gamma$ and each $y \in Y$, γ maps the fibre $p^{-1}(y)$ onto the fibre $p^{-1}\, p\, \gamma\, p^{-1}(y)$. Since the flow π on W is minimal it follows that Γ is transitive, i.e., given $w_1, w_2 \in W$ there is a $\gamma \in \Gamma$ such that $\gamma(w_1) = w_2$. (To see this, we note that by the minimality of the flow on W there exists a generalized sequence $t_i \in T$ such that $\pi_{t_i}(w_1) \to w_2$. By the compactness of Γ, some generalized subsequence converges, say $\pi_{t_i} \to \gamma \in \Gamma$. Clearly one has $\gamma(w_1) = w_2$.)

We next show that p is an open map. Equivalently, we will show that given any $w \in W$ with $y = p(w)$ and a generalized sequence $y_n \to y$, then there exists a generalized sequence $w_n \in W$ such that $w_n \in p^{-1}(y_n)$ and $w_n \to w$. Thus assume w, y and y_n are given as stated. Pick any sequence $\hat{w}_n \in p^{-1}(y_n)$ and choose a subsequence so that $\hat{w}_n \to \hat{w} \in p^{-1}(y)$. By the transitivity of Γ, there exists $\gamma_n \in \Gamma$ such that $\gamma_n(\hat{w}) = \hat{w}_n$. Now choose another subsequence so that $\gamma_n \to \gamma$ in Γ. Then

$\hat{w}_n = \gamma_n(\hat{w}) \to \gamma(\hat{w})$ and therefore $\gamma(\hat{w}) = \hat{w}$. Since γ maps $p^{-1}(y)$ onto itself, there is $v \in p^{-1}(y)$ such that $\gamma(v) = w$. Let $w_n = \gamma_n(v)$. Then $w_n = \gamma_n(v) \to \gamma(v) = w$ and $w_n \in p^{-1}(y_n)$ QED .

Remark 16. The last theorem is a special case of Auslander [5, Theorem 5] .

References

[1] L. Amerio. Soluzioni quasi-periodiche, o limitate di sistemi differenziali nonlineari quasi-periodiche, o limitati 39 (1955), 97-119.

[2] H. Anzai. Ergodic skew-product transformations on the torus. Osaka Math. J. 3 (1951), 83-99.

[3] Z. Artstein. The topological dynamics of an ordinary differential equation. J. Differential Eqs. 23 (1977), 216-223.

[4] Z. Artstein. Topological dynamics of ordinary differential equations and Kurzweil equations. J. Differential Eqs. 23 (1977), 224-243.

[5] J. Auslander. Homomorphisms of minimal transformation groups. Topology 9 (1970), 195-203.

[6] V. Barbu and S.I. Grossman. Asymptotic behavior of linear integro-differential systems. Trans. Amer. Math. Soc. 171 (1972), 277-288.

[7] N.P. Bhatia and O. Hajek. Local Semi-Dynamical Systems. Lecture Notes in Math. No. 90. Springer-Verlag, New York, 1969.

[8] S.P. Diliberto. On systems of ordinary differential equations. Contrib. Theory Nonlinear Oscillations, Vol. I, pp. 1-38. Ann. Math Studies No. 20, Princeton Univ. Press, 1950.

[9] J. Dugundji. Topology. Allyn and Bacon. Boston, 1966.

[10] R. Ellis. Distal transformation groups. Pacific J. Math. 8 (1958), 401-405.

[11] R. Ellis. Lectures on Topological Dynamics. Benjamin. New York, 1969.

[12] R. Ellis and W.H. Gottschalk. Homomorphisms of transformation groups. Trans. Amer. Math. Soc. 94 (1960), 258-271.

[13] J. Favard. Lecons sur les Fonctions Presque-Périodiques. Gauthier-Villars, Paris, 1933.

[14] A.M. Fink. Almost automorphic and almost periodic solutions which minimize functionals. Tohoku Math. J. 20 (1968), 323-332.

[15] A.M. Fink. Semi-separated conditions for almost periodic solutions. J. Differential Eqs. 11 (1972), 245-251.

[16] A.M. Fink. Almost Periodic Differential Equations. Lecture Notes in Math. No. 377. Springer-Verlag, New York, 1974.

[17] H. Furstenberg. The structure of distal flows. Amer. J. Math. 85 (1963), 477-515.

[18] W.H. Gottschalk and G.A. Hedlund. Topological Dynamics. Amer. Math. Soc. Colloquium Publ. 36, 1955.

[19] J. Hale. Functional Differential Equations. Appl. Math. Sciences No. 3. Springer-Verlag, New York, 1971.

[20] J. Kato and Y. Sibuya. Catastrophic deformation of a flow and non-existence of almost periodic solutions. (to appear).

[21] J.L. Kelley. General Topology. Van Nostrand, New York, 1955.

[22] A. Knapp. Distal functions on groups. Trans. Amer. Math. Soc. 128 (1967), 1-40.

[23] J. Kurzweil. Generalized ordinary differential equations and continuous dependence on a parameter. Czech. Math. J. 7 (82) (1957), 418-449.

[24] I.G. Malkin. Stability for persistent disturbances. Priklad. Mat. Mekh. 8 (1944), 327-334.

[25] D. McMahon and T.S. Wu. Homomorphisms in topological dynamics. Trans. Amer. Math. Soc. 217 (1976), 257-270.

[26] R.K. Miller. Almost periodic differential equations as dynamical systems with applications to the existence of a.p. solutions. J. Differential Eqs. 1 (1965), 337-345.

[27] R.K. Miller. Linear Volterra integrodifferential equations as semi-groups. Funk. Ekvacioj 17 (1974), 39-55.

[28] R.K. Miller and G.R. Sell. Volterra Integral Equations and Topological Dynamics. Memoir Amer. Math. Soc. No. 102, 1970.

[29] R.K. Miller and G.R. Sell. Topological dynamics and its relation to integral equations and nonautonomous systems in Dynamical Systems: An International Symposium Vol. I, pp. 223-249. Academic Press, New York, 1976.

[30] V.V. Millionščikov. On the relation... almost periodic coefficients. Diff. Urav. 3 (1967), 2127-2134. (pp. 1106-1109 of English translation).

[31] F. Nakajima. Separation condition and stability properties in almost periodic systems. Tohoku Math. J. 26 (1974), 305-314.

[32] V.V. Nemytskii and V.V. Stepanov. Qualitative Theory of Differential Equations. Princeton Univ. Press, 1960.

[33] W. Parry and P. Walters. Minimal skew-product homeomorphisms and coalescence. Compositio Math. 22 (1970), 283-288.

[34] O. Perron. Über ein Matrixtransformation. Math. Zeit. 32 (1930), 465-473.

[35] R.J. Sacker and G.R. Sell. Skew-product flows, finite extensions of minimal transformation groups and almost periodic differential differential equations. Bull. Amer. Math. Soc. 79 (1973), 802-805.

[36] R.J. Sacker and G.R. Sell. Finite extensions of minimal transformation groups. Trans. Amer. Math. Soc. 190 (1974), 325-334.

[37] R.J. Sacker and G.R. Sell. Existence of dichotomies and invariant splittings for linear differential systems I. J. Differential Eqs. 15 (1974), 429-458.

[38] R.J. Sacker and G.R. Sell. A spectral theory for linear differential systems. (to appear).

[39] R.J. Sacker and G.R. Sell. Linear Differential Systems. (to appear).

[40] G. Seifert. Stability conditions for the existence of almost periodic solutions of almost periodic systems. J. Math. Anal. Appl. 10 (1965), 409-418.

[41] G. Seifert. Almost periodic solutions for almost periodic systems of ordinary differential equations. J. Differential Eqs. 2 (1966), 305-319.

[42] G.R. Sell. Nonautonomous differential equations and topological dynamics I, II. Trans. Amer. Math. Soc. 127 (1967), 241-262, 263-283.

[43] G.R. Sell. Topological Dynamics and Differential Equations. Van Nostrand-Reinhold, London, 1971 .

[44] G.R. Sell. Differential equations without uniqueness and classical, topological dynamics. J. Differential Eqs. 14 (1973), 42-56.

[45] G.R. Sell. Almost Periodicity and Differential Equations. Lecture Notes. Univ. of Minnesota. (to appear).

[46] N. Steenrod. The Topology of Fibre Bundles. Princeton Math. Series No. 14. Princeton Univ. Press, 1951.

[47] W.A. Veech. Almost automorphic functions on groups. Amer. J. Math. 87 (1965), 719-751.

[48] D.R. Wakeman. An application ... differential equations. J. Differential Eqs. 17 (1975), 259-295.

[49] T. Yoshizawa. Asymptotically almost periodic solutions of an almost periodic system. Funk. Ekvacioj 12 (1969), 23-40.

[50] T. Yoshizawa. Stability Theory and the Existence of Periodic Solutions and Almost Periodic Solutions. Appl. Math. Sciences No. 14. Springer-Verlag, New York, 1975.